FROM NEURONS TO SELF-CONSCIOUSNESS

HOW
THE BRAIN
GENERATES
THE MIND

FROM NEURONS TO
SELF-CONSCIOUSNESS

BERNARD KORZENIEWSKI

Humanity
Books

an imprint of Prometheus Books
59 John Glenn Drive, Amherst, New York 14228-2119

Inquiries should be addressed to
Humanity Books
59 John Glenn Drive
Amherst, New York 14228–2119
VOICE: 716–691–0133
FAX: 716–691–0137

WWW.PROMETHEUSBOOKS.COM

15 14 13 12 11   5 4 3 2 1

Library of Congress Cataloging-in-Publication Data

Korzeniewski, Bernard.
    [Od neuronu do samo swiadomosci. English]
    From neurons to self-consciousness : how the brain generates the mind/
by Bernard Korzeniewski.
        p. cm.
    Includes bibliographical references.
    ISBN 978–1–61614–227–8 (pbk. : alk. paper)
    1. Consciousness. 2. Brain. I. Title.
    [DNLM: 1. Brain—physiology. 2. Consciousness. 3. Evolution.
4. Mental Processes. WL 300]

QP411.K67 2010
612.8'233—dc22

2010030735

Printed in the United States of America on acid-free paper

CONTENTS

PREFACE

In this book, I plan to show how consciousness—and self-consciousness in particular—emerges gradually from the operation of elements of physical reality that are devoid of consciousness, as one passes from a single neuron through more and more sophisticated complexes of neurons to the extremely complex neural network that constitutes the human brain. The passage should be understood in two ways. First, it should be seen as the process of evolution of our nervous system from the phase of a few neurons that manage the behavior of the entire organism (some coelenterate remain in this phase until today) to that of a complicated system—the brain. Second, it should be seen as the process of proper organization of neurons into a functional superstructure, possible due to the properties of single neurons in the human brain, that constitutes a completely new quality absent from the physical world and that of living organisms, namely, that of a subjective sphere of psychic phenomena. The process leading to this seems to be extremely enigmatic. This is so because—as we are, to the best of our knowledge, the only creatures endowed with consciousness (at so advanced a level)—the understanding of the essence of being a human constitutes the stake of this game.

My second objective consists in elucidating the implications of the discovered aspects of human brain functioning for the status and the limitations of our worldview and our cognitive abilities. To simplify the matter considerably, I want to pose and justify the thesis that practically all principal philosophical problems can be reduced to neurophysiology.

Therefore, this book cannot be considered as a popular science publication *sensu stricte*. Although it contains information based on the current state of knowledge in the domain of neurophysiology, it is also a presentation of my own, original vision of the human mind (deeply rooted in modern scientific knowledge), while my opinions coincide with those of other authors. The popular mode of presentation is aimed at facilitating the reading of those who are not specialists in the domain of neurophysiology, e.g., physicists, scholars, as well as lawyers, physicians and students who are curious about their world. However, I wrote this book with the conviction that it can be interesting for biologists of various specialties (e.g., neurophysiologists, ethologists, evolutionists) as well as a wide group of philosophers and psychologists, especially those gravitating toward cognitive science. To keep the right balance, I also attempted to separate popularized scientific knowledge from my own opinions.

The nature and genesis of the mind and consciousness remains a mystery to this day, in spite of hundreds of years of philosophical considerations and scientific research. In the texts I have read, nobody has proposed a sufficiently coherent and satisfactory explanation of the very nature of brain functioning and its correlation with a variety of mental phenomena. Any potential generalizations are usually inundated with a plethora of anatomic and physiological details, lacking any integrating conclusion. Therefore, as my reading of multiple texts and considerations allowed me to work out a vision that ordered and explained (at least to myself) many aspects of the topic in question, I decided to present it to a general readership for their appraisal. I aimed at reaching the very core of the issue and to establish possibly a general and synthetic vision of the entire domain, or set of issues, while abstracting from detailed aspects of multiple phenomena, all of which are interesting in themselves, but not related directly to the essence of mind and consciousness, and thus prone to obscure the general overview (I hope professionals can appreciate this approach).

This text focuses on a set of fundamental questions. What is *consciousness* (in itself)? What does it mean to be aware of something? What are different mental objects—such as sensations, thoughts, ideas/concepts, emotions, memories—"present" in our mind, but not existing in the external, material world? What is the difference between the *sensation* generated by a rose and perceived through the sense of vision, and the *concept* of a rose that can be retrieved from the memory at any moment? Where do the "qualitative aspects" of sensations come from? What does the phenomenon of memory consist in? Why do we feel pleasure and displeasure? What is the neurophysiological correlate of *self-consciousness*, i.e., of the consciousness of one's own being conscious, of one's "ego"? What is the relation between consciousness and self-consciousness? Does the former condition the latter? These are just a few examples of the questions this book will address.

The primary message of the book consists of a set of basic postulates. I believe (and here my point of view converges with that presented by Francis Crick in his book *The Astonishing Hypothesis*) that the functioning of the brain is of an emergent character, i.e., it cannot be ascribed to particular brain parts, such as particular neurons or groups of neurons, which means that the entire brain is not simply the sum of its elements. Consequently, I think that consciousness is a derivative of a certain functional spatiotemporal pattern of impulse transmission within the network of neurons. I attempt to show that the representation of the world in the neural network is connotative (relational) in character and not denotative (absolute); it is of fundamental importance for the nature of mental objects and our cognitive abilities. Finally, I propose the thesis that the relation of self-reference/self-application—lying at the base of numerous logical paradoxes—is responsible for the emergence of consciousness.

I would like to express my gratitude to Andrzej Joachimiak for his careful reading of the entire manuscript and encouraging me to remove certain repetitions, which added clarity to the entire text. I am also grateful to Mariusz Papp for a discussion on the reward system. I would like to express my gratitude to all those who challenged me intellectually and thus contributed to this book. However, first of all I would like to offer special thanks to Paweł Wawrzyszko for translating this book.

1.

THE MAIN IDEA

Before embarking on a journey it is good to select a direction, to have a compass with which to constantly check bearings. The distant mountain of our goal must always be kept in view. Without this strong compass we can so easily become waylaid and loose all impetus and motivation.

For this reason, I would like to offer the main focus of this book in order to ensure that readers maintain the sense of direction when getting acquainted with facts, arguments, and concepts; that means recognizing the purpose of their presence at a given point, knowing how to locate them within the framework of the overall vision, and how to interpret them in this context. First of all, I contend that mind and self-consciousness are derivatives of the functioning of the human brain. In particular, I think that each type of mental phenomena—sensations, thoughts, emotions, memories, concepts/ideas—is based on particular spatiotemporal patterns of neural activity of the brain, differing slightly in the case of each specific phenomenon.

In my opinion, the sense of total dissimilarity between the subjective zone of consciousness and the external material world, i.e., the emergence of the psychic level of consciousness, is due not so much to the degree but rather to a specific type of complexity of the functional organization of signal transfer in the neural network. I think that the specific type of complexity leading to the emergence of self-consciousness—the concept of consciousness does not make any sense without the concept of self-consciousness, in my view—consists in the relation of self-focusing (the orienting-on-itself) of the cognitive and decision-making center in the human brain. As a consequence, I think that the attribute of complexity (as well as related phenomena, e.g., information) should be granted the status of an aspect of the world that is as objective as space, time, or matter. I think that the "subjective" zone of mind exists in our world in a perfectly objective manner, just like the phenomenon of life. An organized SYSTEM has to appear for both of these phenomena to come into being. Thus, self-consciousness—as a type of physical complication of the external physical world—is our long-distance goal.

2.

THE FUNCTIONING OF A NEURON

s mentioned above, I maintain the opinion that the subjective mind, the psyche and consciousness are derivatives of the neurophysiological functioning of the brain. Thus, I reject the philosophical view that there exists the spirit or the soul, understood as totally autonomous entities, separate from matter and responsible for adding the "psychic" dimension to the material seat (I do so for numerous reasons, but primarily due to the fact that such approach cannot be reconciled with any rational scientific methodology). To the contrary, I think that the subjective zone of human consciousness is an emergent phenomenon, i.e., it emerges as a result of a certain complex and specific aggregate of neurophysiological processes. This conviction, however, is not equivalent to consequent materialism and reductionism, for I believe that the so-called external world (matter included) accessible to the human mind is shaped to a great extent by our psyche, and that a consequent semantic analysis of the concept of matter leads inevitably to the void of nonsignificance. This is an expression of my moderate episte-mological skepticism, as I believe that the biological evolution of the human brain considerably conditions and limits our cognitive abilities. However, if we refrain from analyzing in an excessively strict manner the sense and justification of the "objective" status of matter, and from ascribing to it the attributes of the absolute, we can maintain (and remain within the boundaries of common sense) that the so-called spirit is a derivative of certain processes that occur in the external world of matter.

If the spirit is to emerge as a result or by-product of a certain specific manner of functioning of matter, there immediately arises the question what is the material basis of the mind or—in more specialised terms—what is the physical correlate of mental processes? We all know perfectly well that the human brain is the only known seat of self-consciousness[1] in the universe (it may be the case that the incipient forms of self-consciousness could be found in apes and dolphins). What is there in the structure and functions of the brain that makes it so exceptional—the only entity capable of "generating" subjective psychic phenomena? The brain consists of the same elementary particles, atoms, and chemical compounds that can be found outside the brain, in inanimate objects or living organisms (in the case of organic compounds). Thus, the specific attributes of the structure and functions of the brain, which are responsible for its psychic dimension, should be sought for at some higher level of the organization of matter. The neuron, with its ability to quickly direct and transmit information signals at large distances, is the first step—specific for the brain—of the ladder of successive degrees of complexity. The functioning of particular components of the neuron can be explained in a satisfactory manner at the physical, chemical, and biochemical levels. There is nothing here that could elucidate the essence of the phenomenon of consciousness (although the biochemical and biophysical aspects of the functioning of particular neuron components are delightfully purposeful and "ingenious"). Therefore, at least the entire neuron should become the starting point in the quest for the origins of the mind. Neurons can of course be found in animal brains as well, the majority of which do not show signs of psychic activity, and in primitive neural networks of worms, for example, where one cannot even speak of the brain. This fact implies that the seat of consciousness cannot be found in the neuron alone, but in a network of (very) many neurons organized in a certain manner. To understand the functioning of such a network, however, one has to begin with the specific features of the functioning of single neurons. Therefore, we have to start our journey from the individual neuron to reach our destination, i.e., self-consciousness.

The neuron is one of multiple types of animal cells and, as such, it contains the typical components: the nucleus, cell membrane, cytoplasm with cytoskeleton (e.g., microtubules) and a number of organelle like mitochondria, lysosomes, endoplasmic reticulum, etc. On the other hand, neurons differ considerably from other cell types with respect to both their structure and their functions. During the process of biological evolution neurons specialized to realize a single primary task, namely, the quick and strictly directed transfer of signals at large distances. That is why their structure is so characteristic: each cell is equipped with very long, branching processes (an axon and dendrites, discussed below in greater detail) that transmit electrical impulses. Neurons receive impulses from other neurons or receptor cells (e.g., light-sensitive cells in the retina of an eye) and may (although they do not have to) transfer impulses further to other neurons or effector cells (e.g., muscle cells). However, the communication among neurons and between neurons and receptor or effector cells is not the only type of transmission of information between cells in an animal organism. On the contrary, there are many cases where cells transmit information to their neighbors by secreting chemical compounds into the intracellular space (this is the so-called paracrinic transfer of signals between cells). This type of communication, however, operates only at very short distances. A more extensive type of communication (covering the entire body) is based on specialized groups of cells (glands) that secrete signalling substances (hormones) into the blood system (this is called the endocrine manner of signal transmission). Such transmission is potentially directed at all cells in an organism. Yet, only some cells—those equipped with appropriate receptors in the cell membrane—react to a given hormone and the type of generated reaction depends on the type of reacting cell. Such signals need a rather long time to be transmitted (from the moment of secreting a hormone into the blood system until the moment of reaction of a cell "sensitized" to a given hormone), because it is limited by the speed of blood circulation. Neurons, like other types of cells, also transmit signals chemically, through neurotransmitters secreted into the synaptic cleft (to be discussed in greater detail later on). In contrast to the paracrinic and hormonal communication, neurons transmit impulses very quickly and at long distances to strictly defined target cells.

Why these specific characteristics of neurons and why is the nervous system built of these cells so important for the emergence of the mind? Neural communication has developed during biological evolution from simple chemical communication between neighboring cells (the above-mentioned paracrinic signal transmission). The neural transfer of information differs dramatically from this type of communication and it is related to the highly specialized structure and functions of neurons. The neural transfer provides a totally new quality that allows for the coordination of the behavior of multicellular animal organisms. The properties of neurons allowed them to become building blocks of highly complex systems for information processing. Elements of such systems must very quickly transmit signals at long distances and direct them to specific targets so that the entire system can be properly integrated and operate accurately and reliably in real time. I believe that only such an information processing system can potentially become the "seat" of the mind. Thus, it is this huge potential gathered in the neural communication that led (after several hundred million years of evolution, after the emergence of the first multicellular animals equipped with a nervous system) to the emergence of the mind and self-consciousness. Let us consider the structure and functioning of a typical nerve cell.

Figure 1 is a schematic representation of a neuron. A neuron consists of a more or less spherical cell body containing the nucleus and of very long and highly ramified processes (dendrites and axons) responsible for the transmission of electric impulses. Numerous dendrites lead impulses to the soma of a cell (from other neurons or receptor cells), while a single axon transfers impulses further on (to other neurons or effector cells). A highly ramified axon end of the preceding neuron along the information-transfer route communicates with equally ramified dendrites of the following neurons through synapses (of course, each neuron receives signals from numerous other neurons and transmits them to numerous other neurons). A synapse consists of a mushroom-shaped bud of an axon ramification (the presynaptic part) pressed against a mushroom-shaped bud of a dendrite ramification (the postsynaptic part) that are divided by the synaptic cleft. The latter divides physically the presynaptic (axon) part and the postsynaptic (dendrite) part, and allows them to communicate only through chemical means.

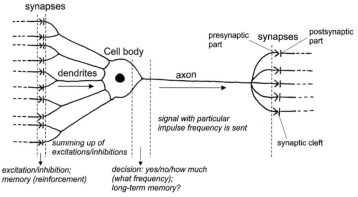

Fig. 1. The general scheme of a neuron. The following elements are presented: the cell body, impulse-conducting processes (dendrites and the axon) and synaptic connections between neurons. The figure also describes the functions of particular neuron elements in the process of conducting nerve impulses.

Nerve impulses are transmitted along the processes of neural cells. In the case of axons, the process is active and consists of an impulse moving quickly from the axon hillock to axon endings. The impulse takes the form of a wave of depolarization of the cellular membrane (the resting potential is replaced by the so-called action potential) that involves a redistribution of Na^+ and K^+ ions across the membrane. The quantitative information contained in nerve signals is coded as the frequency of reappearance of particular impulses (that can vary from 30 to 1000 impulses per second) and not as the intensity (the action potential level) of the impulses that remains constant. In dendrites, impulses are conducted passively, which involves a general change in the membrane potential. An impulse "jumps" from the axon of the preceding neuron to a dendrite of the following neuron through the synaptic cleft. The presynaptic part of a synapse contains vesicles with a chemical compound called a neurotransmitter. When a nerve impulse reaches a synapse the content of the vesicles is released into the synaptic cleft and the neurotransmitter molecules bind to the corresponding protein receptors in the postsynaptic membrane, which leads to a change in the electric potential of the dendrite connected to the synapse that propagates to the cell body. The amount of the neurotransmitter appearing in the synaptic cleft is proportional to the intensity of the signal (the frequency of impulses arriving to the presynaptic part). A whole range of neurotransmitters has been discovered that includes glutamate (the main neurotransmitter in the brain), GABA (appearing at inhibitory synapses), acetylcholine (participating in the establishment of memory traces and stimulating muscle contractions), noradrenaline and serotonin (controlling the general excitation and the "mood" of the brain) and dopamine (involved in the "reward system"). There also exist rare electric synapses independent on chemical signalling that are capable of carrying an electric impulse directly over the synaptic cleft.

It is not necessary here to discuss the technical (physicochemical and biochemical) details of both active impulse transmission along axons and the chemical communication in the synaptic cleft, as they are not directly related (in my opinion) to the origins of consciousness. For it is the general manner in which neurons process information, and not their constructive material or detailed mechanisms responsible for information processing, that is important for the emergence of the mind. If neurons and their interconnections were substituted by an adequate number of digital elements (made of metals, silicon, etc.) interconnected in the same way and operating like neurons (as far as cybernetic and information processing aspects are concerned), such a system would allow consciousness to emerge. (The possibility of "inducing" subjective mental states in digital devices will be discussed in greater detail in chapter 8 devoted to artificial intelligence.)

In these considerations, it is crucial that neurons and their processes are not simple cables conducting impulses. A typical neuron connects through synapses with several thousand other neurons. If each stimulated nerve cell stimulated, for instance, five thousand other nerve cells, all neurons in a brain would quickly reach the maximum level stimulation that would paralyze the operation of the brain. Of course, brain neurons do not function like that.[2] First of all, each synapse has a characteristic threshold potential, i.e., the minimum intensity of the signal (impulse frequency) that activates a given synapse. Second, apart from excitatory synapses (and neurons) there are also inhibitory synapses (neurons). Once the excitation threshold is exceeded, they reduce the electric potential in the target neuron (its dendrite and the body of the cell) to the level that can even be lower than the original value. Thus, their operation counterbalances the effects produced by excitatory synapses. In both cases, the value of the excitation threshold of a given synapse determines the sensitivity of the synapse to incoming signals: the lower the threshold, the higher the sensitivity. Thus, the synapse decides—in a certain elementary sense—which impulses go through and which are stopped. A neural route, a neural circuit or a functional "object" in a neural network (all of them consisting of numerous neurons) are "channeled" or "blazed" (signals easily take a given route) if the threshold potential of excitatory synapses is low, while that of inhibitory synapses is high. This is similar to the behavior of water: it flows primarily through grooves on a flat surface of a horizontal rock and fills in the hollows, while only small amounts flow over the surface of high "plateaus."

This is not all, however. Some synapses are able to make more complicated decisions. Such capability is observed in synapses equipped with the so-called NMDA channels in the postsynaptic membrane. To avoid technicalities, I would only like to indicate that such synapse responds not only to the presence of a neurotransmitter in the synaptic cleft, but also to the level of the electric potential at the dendrite end (and therefore in the postsynaptic part of the synapse as well). If the potential rises as a result of "passing" a signal through a neighboring synapse (or synapses), the synapse in question may be activated by a much less intense signal (much lower neurotransmitter concentration) than in the case of the passive potential (i.e., the heightened potential lowers the excitation threshold of the cell). In other words, the synapse in question will be activated (it will transmit the signal) only when it receives a certain amount of a neurotransmitter, while at least one of its neighboring synapses is also activated. This is a typical case of a logical conjunction: "if A and B, then C," where A refers to the presence of a neurotransmitter, B refers to the heightened potential (due to the activation of a neighboring synapse), and C refers to the activation of the synapse itself. The electric potential, in turn, can be lowered by the neighboring inhibitory synapses (releasing Cl^- ions into the neuron), which deactivates the synapse equipped with NMDA channels. It can be seen that quite complex decisions are made at the level of synapses that decide whether to pass a neural signal over or not.

Much more complex decisions are made in the body of a nerve cell or, to be exact, in this small area where the axon leaves the cell. As distal ramifications of dendrites converge toward the cell body, the levels of excitatory (and inhibitory) signals reaching particular synapses of these dendrites are summed up (a part of synaptic connections is located directly over the cell body). The final summation is carried out at the root of the axon (a certain electric potential value is established). It is simply the result of electric potentials originating from the postsynaptic endings of a given neuron. Depending on the resultant value, a neuron decides whether signals are sent through the axon or not, and if they are sent, what frequency they will have. The generated signal reaches all synaptic endings of the axon of a given nerve cell and the decision-making process starts all over again in the implicated neurons that figure out what to do with the received signals (from the neuron in question and from other neurons).

Neurons may not only process (in a complex manner) the information that has reached them. They may also modify the complex "logical function" that they carry out and do so on the basis of their past experience (excitations) by changing the "weight" of the synaptic connections inversely proportional to the value of the excitation threshold. An activated synapse remains more sensitive to signals for some time (its excitation threshold is lowered). In ordinary conditions, the increased weight of a synaptic connection disappears after some time and the excitation threshold returns to its original value (that corresponds to short-term memory). Things are different in the case of the so-called long-term synaptic potentiation that occurs primarily in the already-mentioned synapses equipped with NMDA channels. The activation of such a synapse results in the release of Ca^{2+} ions into the postsynaptic ending that, in turn, initiates a cascade of biochemical processes that lead to a permanent increase in the weight of a given synaptic connection. This is one of the basic mechanisms of long-term memory and learning. The establishment of new synaptic connections or elimination of existing ones are other examples of the modification of the functions of a nerve cell. It is supposed that the body of a nerve cell may also participate in the formation of memory traces, for instance through the regulation occurring at the genetic level, which leads to a change in the sensitivity of the axon root to potential increases or to the production of macromolecules that are carriers of at least a part of memory traces.

Let us sum up the above considerations. In general, the primary feature that distinguishes a neuron from other cells is its ability to transmit signals carrying biologically relevant information:

a) at long distances. This is ensured by the shape of a neuron, and in particular by the length of its processes, i.e., dendrites and the axon. This allows different brain regions to communicate easily with each other.

b) very quickly. This is ensured by the special structure of a neuron, e.g., by the electrically insulating myelin sheath of the axon. Thus, the brain can process huge amounts of information and react in real time to different events in the external world. It may also model these events.

c) to strictly defined receivers. Particular axon ramifications connect (through synapses) to dendrites (or cell bodies) of particular target neurons. This ensures the possibility of an extraordinarily complex information network that realizes highly complex and biologically purposeful tasks. As mentioned before, an average neuron communicates with several thousand of other neurons. Since the human brain contains about a hundred milliard of neurons, the number of possible combinations of interconnections between them exceeds the number of atoms in the observable universe by many orders of magnitude.

More, a neuron is by no means like a simple industrial cable. As mentioned above, both in the synapses and in the nerve cell body, decisions are made concerning the generation of a signal in response to an excitation and the intensity of this signal (the frequency of impulses). The "decision-making criteria" (i.e., the excitation threshold or the connection weight) are not strictly defined and may be modified as a result of past experience (activations of a given synapse) as well as (or rather: in particular) due to the temporal coincidence of excitation of several neighboring synapses. This is the basis for learning and memory. Neurons are information-processing systems that carry out complex logical operations. However, they do not follow a discrete logic (and certainly not a zero-one logic) that operates on sharply defined symbols and rules that determine the symbol interrelations as in the case of contemporary computers. To the contrary, neurons are characterized by analog and continuous information processing. The signal strength (impulse frequency) changes in a continuous manner, which applies also to the summation (at the axon root) of potentials originating from different dendrites and their corresponding synapses. It can be said that the logic of neurons is, to a large extent, imprecise and approximate, and it operates on fuzzy sets (and symbols). As such, it is not quite suitable for carrying out fast and perfectly precise mathematical calculations, which computers excel in. Yet this is the feature (in my opinion) that is the absolutely necessary condition for the emergence of the psyche and consciousness in the human brain (or other appropriately structured and appropriately functioning object). Without this feature, the emergence of a relationally organized conceptual network that signifies by connotation would be impossible. And this network constitutes the substance of our psyche as I propose in chapter 7.

3.

BRAIN STRUCTURE AND FUNCTION

eurons are building blocks of the entire nervous system, including the human brain that is the carrier of consciousness.[1] What is the nervous system? The main "sense" of its existence is found in the inputs and outputs, i.e., the receptors and effectors, because the primary biological function of neural networks consists in receiving a stimulus (representing a feature of the external world or an event occurring in this world) and "transforming" it into a behavioral pattern of the organism (an adequate reaction to the stimulus). This elementary role of the nervous system is expressed in the simplest form in certain cnidarians equipped with only a few nerve cells (or quasi-neurons). These cells receive signals from the receptors sensitive to mechanical irritation and, if the signal is sufficiently strong (supraliminal), they activate muscle cells (epithelial muscle cells) along the sea anemone body. As a result, the sea anemone contracts, which can protect it from a predator.

The operation of the human brain is in a sense just a surplus to the execution of this simple function. During the simultaneous evolution of the systems of receptors and effectors and of the nervous system itself interconnecting the former two systems, the operation of transforming received stimuli into appropriate behavioral patterns became much more complex and highly mediated (the external world was being mapped into the network of nerve cells in a constantly more accurate manner). It is not possible to understand the basic principles of brain functioning and especially the cognitive aspect of its functioning (that is, to my mind, most closely related to the genesis of consciousness) without being aware of this biologically primordial function of the nervous system of living organisms.

Before considering the structure and functioning of the human brain, I would like to make a certain qualification. This book focuses on such issues as the cognitive functions of the brain, the establishment of a certain representation or model of the external world in the brain, the functioning of the "operating memory," the processes of decision making and activation of effectors (mainly muscles) and coordination of their activities. In short, I attempt to describe how the above-described function of transforming environmental stimuli into animal behavioral patterns was improved (and mediated) during biological evolution leading from a few neurons of a cnidarian to the human brain and how self-consciousness—which is, in my opinion, only a by-product of the process of improving the cognitive brain functions based on the simple sequence: stimulus → processing → response—appeared on the scene (or rather entered it by the backdoor). I will also discuss the neurophysiological foundations of emotions because they also have an important position in our consciousness. However, I will skip many other functions of the brain that are essential for survival, but not directly related to the genesis of consciousness, such as the control of hunger/satiety feeling achieved by the regulation of the glucose level in blood or the regulation of day and night activity of an organism (the sleep and wakefulness cycle). Those interested in these issues will find more information in a number of popularized or professional works on these subjects.

For the same reason, I will skip the anatomical details of the brain structure that are derivatives of so many of its functions and of its biological history. The brain structure is, to a large extent, a result of the events along the evolutionary path that has produced the extraordinary complexity of various structures resembling the layers and geological forms oftentimes deformed due to tectonic shifts of the lithosphere. When viewed from outside or presented as a cross-section, the brain seems hopelessly entangled, consisting of various winding structures and centers (it is sufficient to take a look at any schematic of brain anatomy to confirm the impression of complexity). The names of different brain components (fanciful and highly varied) indicate both the complexity of the brain anatomy and our disorientation regarding this plethora of forms. It suffices to mention such terms as: cingulate gyrus, formatio reticularis, locus coeruleus, thalamus and hypothalamus, caudate nucleus, globus pallidus, medial geniculate nucleus, central sulcus or mamilliary body. I will attempt to show, however, that there is a method in all this madness, and I will do so by isolating the general principles of brain functioning (primarily the cognitive aspects) and showing that they can be rather clearly explained without reference to the entire collection of brain structures, since just a few anatomic terms suffice to do so. It will be necessary to abstract from numerous details of the brain structure and functions, important as they are, and to focus on the most important aspects of brain functioning. Thus, particular trees will not prevent us from seeing the entire forest.

Let us return to the interrupted considerations of information transfer along the route receptors—nervous system (data processing)—effectors. I will attempt to present a general outline of the brain prepared for these considerations, and to show the essential difference between the middle part of the scheme in the case of the brain, on the one hand, and in the case of a few nerve cells of a cnidarian, on the other hand. Figure 2 presents a quite simplified set of elements of the sensorial/cogni-

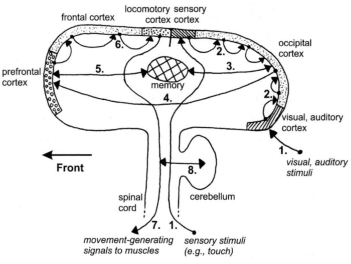

Fig. 2. The general diagram of the human brain, reduced to the most important elements. The diagram focuses on the system that collects external information, construes an operating model of the external world, decides which reactions to stimuli are correct, and triggers the operation of appropriate muscle groups. 1. Stimuli from receptors. 2. Hierarchically organized integration (in numerous stages) of sensory perceptions. 3. Confrontation of sensory perceptions with the existing memory records and formation of new memory records. 4. The sensory-locomotor tract that feeds integrated sensory perceptions into decision-making centers. 5. Confrontation of the operating memory (the decision-making center) with the existing memory records and formation of new memory records. 6. Transcription of general decisions into detailed directives for particular effectors. 7. Signals that activate particular effectors. 8. Movement coordination and controlling feedback.

tive/decision-making/executory "sub-assembly" of the human brain (decidedly the most important one in the context of the psyche and consciousness). The first thing that focuses our attention is a certain symmetry between the sensory part (sensory-cognitive) located in the back of the brain and the locomotor part (decision-making and executory) located in the front of the brain. Signals from sensory end organs (indicated with 1 in figure 2) are transmitted to the sensory cortex. Impulses from tactile receptors, temperature and pain receptors, etc.—located over the surface of the entire body—travel to the brain through the spinal cord and reach the brain and the sensory cortex located behind the central sulcus that divides the frontal (motor) lobe of the cerebral cortex from its parietal (sensory) lobe. In turn, the nerve connections from eyes and ears lead, respectively, to the visual cortex located in the occipital lobe of the cerebral cortex and the auditory cortex located in the temporal lobe (as figure 2 presents a cross-section of the brain that does not go through the temporal lobe, the auditory cortex was arbitrarily moved to the back and attached to the visual cortex). Information reaching a receptor (e.g., the retina) can already be partially processed and integrated within this receptor. Certain features or categories (e.g., spots, colors) are extracted from the information flow and they become the "content" or "substance" of sensory perception. However, the essential part of the signal integration process is carried out in the cerebral cortex. This is where the most complex and general qualities and properties are determined and "organized" into complex objects and relations, and this is where different objects, persons, events, and meanings (e.g., of linguistic utterances) are recognized. It is interesting that the further away from the sensory cortex (visual, auditory, or sensory) one gets, and the deeper down into the area between them, into the parietal and temporal cortex, the more complex and general qualities, objects, and dependencies appear as a result of progressive integration of sensory data. In other words, in the sensory cortex (located in the back part of the brain cortex) it is possible to identify a gradient of increasing hierarchy of generality and complexity of qualities that are extracted from the received sensory data (figure 2 shows a number of arrows marked with digit 2). At each stage of sensory data processing (i.e., at each hierarchy level), there occurs interaction with the memory records in the brain (they are presented in figure 2 as a small spot, but in reality memory—in its different forms—is dispersed practically over the entire area of the brain cortex). The interaction is bilateral. On the one hand, the signals received from receptors can be recorded (at different stages of integration), i.e., they may modify the already existing memory records. On the other hand, sensory stimuli are interpreted, integrated, and understood precisely through their reference to the already-existing memory. Several types of memory can be differentiated, e.g., semantic memory (memory of meanings) that allows one to determine (understand) "what is what" and contains general rules and categories; episodic memory concerning particular objects and events; procedural memory responsible for learned skills such as bicycle riding. Probably, there are no absolute borders between different types of memory. While the brain structure called hippocampus seems to play an important role in the creation of at least some types of memory records, the records themselves are most probably dispersed over large brain cortex areas. The confrontation of integrated sensory data with memory records is marked with number 3 in figure 2.

The process of integration of signals flowing in from receptors finally terminates, of course. Ultimately, the pieces of information coming from different sense organs must be somehow interconnected (e.g., the sensation of the sea is not only the view of large extent of water, but also the sound of waves, the feeling of fresh breeze on the skin, the characteristic smell, etc.), and the integration process must reach the highest level in the hierarchy that contains already fully shaped objects and processes in the neural network corresponding to sensations. This final integration is reached somewhere in the middle areas of the parietal and temporal lobes. What happens next? This part of the sensory cortex is connected by a large tract of nerve connections (axons)—indicated with digit 4 in figure 2—to the front part of the brain cortex responsible for decision-making and locomotor functions, namely, to the prefrontal cortex, where the decision-making center of the brain is located. It is supposed that this is the location of the "operating memory" (by analogy to the operating memory of a computer) that analyzes on-line the integrated sensory data flowing in through the above-mentioned bus and takes decisions concerning current activities on this basis. This is also the place where different objects and events get associated with each other (therefore it is also referred to as the "associative cortex"), compared to memory records, generalized into categories and rules governing the course of events. Finally, this part of cortex makes predictions, formulates long-range plans, and carries out the autonomous activity aimed at analyzing and processing the recently obtained and formerly stored information, i.e., what can in rough terms be called the process of thinking. The operating memory is closely related to the so-called short-term memory. In man, its content, or at least a part of it, can become the content of consciousness (this will be discussed later on). Some elements of the short-term memory are recorded in the long-term memory. When a decision is made (the "resolution" of different, oftentimes competing, groups of neurons) concerning an action (e.g., the pursuit of prey), it has to be put into practice, orchestrated through appropriate "executive acts," i.e., instructions that activate the relevant effectors (primarily muscles) in the relevant temporal sequence. Information processing here is also hierarchical and occurs by stages: the general directives are gradually "transcribed" into more and more detailed instructions concerning contraction of appropriate groups of muscles and particular muscle fibers. All this occurs in the frontal cortex located behind the prefrontal cortex, and later on—in the motor cortex bordering (across the central sulcus) on the already-discussed sensory cortex. The process of transcribing decisions concerning movement into a spatiotemporal pattern of single muscle stimulation is marked with the number 6 in figure 2. It should be noted that the gradient of generality here has the opposite direction to the gradient of sensory data integration: it goes from general and complex to more detailed and elementary dynamic processes in the neural network. Both thinking, making decisions, and transcribing them into stimulations of particular effectors at different hierarchical levels are entangled into the already-existing memory records, and they can participate in the creation of new records (as it occurred in the case of sensory data integration). This process is marked with the number 5 in figure 2.

Finally, the motor cortex emits signals that stimulate particular muscles. Nerve connections leading to muscles leave the brain through the spinal cord (number 7 in figure 2) and spread over the entire body. The coordination of contractions of particular muscles and the control of muscle tension are realized by the cerebellum (8)—the center of complex learned skills (e.g., driving a car).

In principle, this is the most general description of the cognitive and decision-making functioning of the human brain. Before I get on to a more detailed analysis of particular stages of the discussed process, I would like to sum up the above. As already mentioned, and visualized in figure 2, the brain (cortex) is characterized by a quite clear symmetry (both anatomic and functional) between sensory and motor parts. While the sensory-cognitive part constitutes the "ascending" path for nerve signals (this is the path of gradual generalization and growing complexity of "neural objects" and their corresponding "mental objects," i.e., perceptions and concepts that integrate single signals from receptors), the motor and decision-making part constitutes the "descending" path (abstract and general thought processes lead to unanimous decisions governing the behavior of the organism that are in turn concretized, elaborated, and transcribed into stimulations of particular muscles, like the score of a symphony is orchestrated into the voices of particular instruments). Memory plays a very important role at each stage of integration and transcription. During these processes, new memory records are formed, but the existing records also shape the process of integration and transcription. Memory covers a variety of stages of consolidation/transcription. It stores both simple and complex qualities, objects, events, decisions, and locomotor skills.

It is worth adding here that the parts of the brain cortex that carry out the processes (both of integration and transcription) at low hierarchic levels of complexity, i.e., the senses-related cortex (sensory, auditory, and "lower" areas of visual cortex) in the sensory-cognitive (back) part and the locomotor cortex in the locomotor and decision-making (front) part are relatively old in the evolutionary sense. The cortex areas that perform the processes of the highest degree of integration and complexity, i.e., the part of the parietal and temporal cortex where the final stages of sensory data integration occur, and the prefrontal cortex—the "carrier" of the operating memory and thought processes (and the decision-making center) constitute the most recent evolutionary acquisition. These areas—the prefrontal cortex in particular—are best developed in man (compared to other animals, including apes and other primates).

The sensory-locomotor tract transmits highly integrated "sensory images" from the parietal and temporal cortex to the prefrontal cortex, where they reach the current operating memory and appropriately modify the partly autonomous processes of thinking, planning, and decision making. Simplifying the matter slightly, one can say that the image of the external world—both the one received at every moment (properly integrated perceptions) and the one fixed in memory records—is located in the parietal/temporal cortex, while the prefrontal cortex houses the processes (e.g., thinking, association, planning, and the sense of one's own "ego") that are considered to be the internal phenomena of the human psyche. While the mentioned tract constitutes the connection between the two primary functional subsystems of the brain cortex. As the image of the external world is spatially separated in our brain from the "operating center" of the psyche—probably due to evolutionary incidents—the function of the sensory-locomotor tracts consists in remedying this structural imperfection. One can suspect, however, that this task is not perfectly realizable and that the communication between the sensory-cognitive and locomotor-decision-making parts of the brain is not so perfect (due to the time lapse in signal transfer for the distance of over 10 cm between the back and the front part of the brain, if not for other reasons) as it would have been, had the two parts been in direct contact. It seems quite probable that this "technological failure" has an immense impact on our world vision, including our philosophy (this possibility will be considered in more detail in chapter 9). It may be the case that this feature is responsible for the dramatic division in our mind between the spirit (the subjective mind, the psyche) and matter (the external world, the objective reality). In general, I believe (and I discuss this issue further on) that a great majority of our philosophy—ontology and epistemology in particular—results directly from the features (oftentimes shaped by evolutionary incidents) of our human neurophysiology.

Could our brain have been constructed in a better way? In my opinion, of course it could! Let us consider figure 3 that compares a very general diagram of the structure and functions of the human brain (presented in the convention adopted in figure 2) to an imaginary "structurally correct" brain. In the human brain, the parts of the "low order" of the ascending and descending paths border on each other, while the "high order" centers are spatially separated, which creates the need for a connecting tract. In the "structurally correct" mind, however, the high order centers of the ascending path border directly on the high order centers of the descending path, which eliminates the

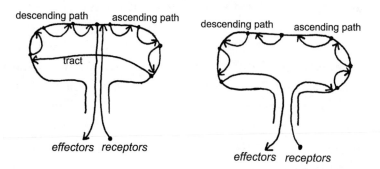

Fig. 3. The ascending and descending paths of information processing in the human brain and in the "structurally correct brain." It is worth noticing that an optimally (economically) constructed brain would not need the sensory-locomotor tract.

need for the tract. Of course, the latter solution is structurally much better. Not only does it avoid wasting the material (very long axons) for the unnecessary tract (it should be noted that the entire length of nerve connections in the human brain exceeds considerably the length of connections in the "structurally correct" brain—see figure 3), but also considerably reduces the time of impulse transmission between the sensory and the loco-motor part of the brain cortex. The structural imperfection of the human brain is of course due to the cumulative character of the biological evolution. In the past, the embryogenesis of the brain was accidentally shaped in such a manner that the higher-order parts of both sensory and locomotor cortex were added—during the process of evolution—outside of the evolutionarily older lower-order parts, and not introduced inside, in between the already-existing parts of the cortex. We have become so accustomed to viewing the human brain as the highest and the most complex of all known creations of nature (which is true anyway) that we forget the fact that some of its structural aspects are rather shoddy. (On the other hand, however, it is worth inquiring, whether man's psyche and his sense of "ego" would have at all come into existence if the brain were struc-turally "optimal." Would there arise philosophy, religion, and science?) The same can be observed in the case of the retina (the light-sensitive part of the eye) in man (and other vertebrates) that is also, formally speaking, a part of the brain delegated for special purposes—visual stimulus recording—during the process of embryogenesis. The retina of vertebrates is also an example of an obviously defective construction. Nerve connec-tions go from the light-sensitive cells (rods and cones) toward the inside of the eyeball, crossing the path of light rays. More, they impose the existence of the so-called blind spot (well known to psychologists) at the place where they concentrate to form the optic nerve that pierces the retina to reach the brain—this is a completely unused fragment of the retina. It is not necessary to invent imaginary examples to prove that a better solution could have been found here, as cephalopods (e.g., the octopus) found it. Their eye is constructed in a similar manner as the eye of vertebrates, but the nerve connections go from the light-sensitive cells to the peripheries of the eyeball. Therefore, even if we can consider man as the best developed biological species, he does not constitute the best solution in every respect.

At the end of this (highly digressive) summary, it should be underscored that the brain does not contain any "super-brain," any "core" of the psyche, any center that constitutes the seat of self-consciousness. If it did exist, a damage of this part would cause the total disappearance of consciousness. On the contrary, damages of different parts of the brain may impair different aspects of the psyche. It proves that the psyche and consciousness are de-localized and they are constituted as derivatives of the interlinking various neurophysiological processes, rather than of a single process.

Let us consider now in more detail the functioning of the ascending and descending paths in the brain. The general diagram of information flow and processing in the human nervous system is summed up in figure 4. The physical stimuli from the environment are registered by receptors and transformed in the process of integration into complex "sensory images" that are further transferred to the decision-making center (the operating memory). They are used there in the decision-making process that, in turn, leads to activating appropriate effectors once the decisions are transcribed into detailed instructions. Both in the process of perception integration and in that of decision making the crucial part is played by the memory records that can be enriched and modified as a result of these processes.

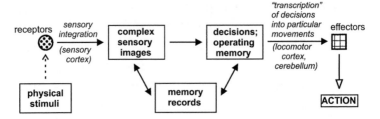

Fig. 4. The general diagram of information flow through the human brain, demonstrating the processes of sensory stimuli integration, the transcription of decisions into muscle movements, as well as the memory records participating in both aforementioned processes through the relevant feedback.

It is worth noting that, formally speaking, the entire nervous system of higher animals (including humans) realizes the same data transfer scheme—stimulus → processing → response—that was discussed in the case of the cnidarian's few nerve cells. What has the obvious progress consisted in? First, it consisted in a truly tremendous increase of the degree of complexity of the central segment in the above information transfer route, namely, that of information processing. Increased amount, diversity, and precision of receptors as well as the development of the sensory-cognitive part of the brain have allowed it to consider much more differentiated aspects of the external world, while the development and improvements in the system of effectors (primarily complexes of muscles, including the hand manipulation system) and the locomotor-decision-making part of the brain have considerably increased the diversity and adequacy of reaction to various environmental stimuli. Second, the function of transforming environmental stimuli into organism behavior has been highly mediated, broken into multiple intermediate stages. Third, information processing ceased to be a simply mechanical, strictly determined process (that is related to the previous point). A system so complex as the human brain cannot operate in a fully deterministic manner. It has to have a random, chaotic component that allows the tiniest disturbance to turn the future evolution of the system in utterly unpredictable directions.[2] Fourth, the entire decision-making system reached considerable autonomy, as it is neither totally nor even predominantly conditioned by the sensory data. It analyzes past experiences (which does not have to bring about results that are applicable here and now) and plans future actions as well as demonstrates a "cognitive inquisitiveness," i.e., the tendency to create a possibly complete vision of the world, the vision that does not bring immediate advantages, but constitutes—at least temporarily—a goal in itself (which does not mean that it is not going to be useful in the future). Last, the human brain has generated self-consciousness, the subjective zone of mental experiences that certainly cannot be attributed to a few neurons of the cnidarian. I will discuss the emergence of the phenomenon of self-consciousness in chapter 7.

Now, we will take a close look at the hierarchically organized integration of sensory data, exemplified by the sense of vision. We will be primarily interested in what happens with the information flowing from the eyes at the successive stages of its processing, and not in which parts of the brain (the visual cortex in particular) carry out the processing. Figure 5 presents a general diagram of visual data processing. The entire process commences, when particular light-sensitive cells of the retina (rods and cones) are stimulated by quanta of electromagnetic radiation within the visual spectrum. The input consists of activated to a varying extent single image points that resemble photographic emulsion grains. It would be gravely erroneous, however, to think that such a "photograph" reaches the highest order centers in the human brain and appears before the light of our consciousness. The information contained in the primary photographic image is considerably transformed at numerous successive stages of processing and correlated to huge information resources already collected in our brain (due to the lifelong process of learning)—otherwise, we would be unable to understand such an image, in much the same way that a camera does not understand the image it records.

The first stages of information processing are already carried out in the retina itself. This is somehow obvious, as the retina contains about a hundred million rods and cones, while the optic nerve contains only about a million nerve fibers (axons), so it is impossible for each light-sensitive cell to independently transfer the information about its state to the brain. The process proceeds in a different manner: the brain receives information that is collected from multiple receptor cells and partially processed. Optic nerves consist of axons of the so-called ganglion cells separated from the light-sensitive cells with a developed network of other neurons. What does the initial information processing consist in? In order to make it possible for *colors* to be recognized, it is necessary to compare signals from several neighboring cones of different types (sensitive to different bands of electromagnetic radiation[3]). "Calculations" of this type are carried out already in the retina by small ganglion cells that ensure high picture resolution (they gather stimuli from small retina areas). A system of other (bigger) ganglion cells gathers signals from both rods and cones. It is "color-blind" and its resolution is low (it collects stimuli from large retina areas), but it operates at a higher speed and is focused on detecting light intensity changes, both in space, i.e., in the retina surface (contrast detection) and in time (movement detection). Thus, it can be seen that the retina does not inform the brain in a univocal and precise manner about everything it records in the visual field. The image transmitted by the retina to the brain is

Image integration

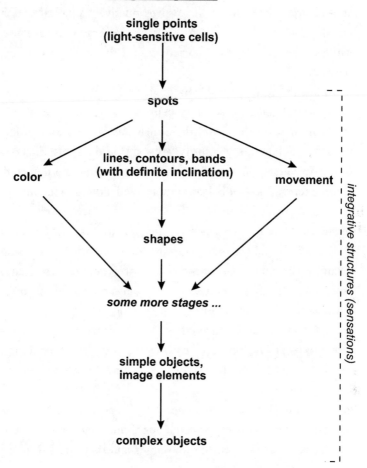

Fig. 5. Particular stages of the hierarchically organized integration of sensory data, illustrated with the process of integration of visual stimuli. This is an example of the ascending (generalizing) path that leads from simple to more complex neural structures.

already considerably distorted. For instance, when the perceived area is large and uniform, it receives relatively small attention and only weak differentiating signals are generated in response, while in the case of more varied stimuli the reaction is much stronger.

Thus the retina transforms single image points (stimulated or nonstimulated light-sensitive cells) into slightly bigger, partly overlapping spots that cover the entire area of the retina. Additionally, an elementary color is "ascribed" to the spots, the contrast between neighboring spots and their movement is detected. Consequently, the "retinal computer" itself, although it contains a relatively small number of nerve cells, carries out quite advanced "calculations" before the visual information even reaches the brain! Further stages of the information processing occur in particular fields of the visual cortex that represent successive stages of a hierarchically organized process of integrating receptor-generated signals. Each of the fields (or at least the fields known as V1 through V4) contains a topological representation of the retina (the so-called retinotopy mapping), which means that neurons receiving signals from neighboring areas in the retina are located one beside the other in the visual fields as well. However, the higher the hierarchy level, the larger the retina area that is supported by particular neurons—i.e., the "receptive fields" of neurons are larger and they react to more complex qualities.

Different image properties at different hierarchy levels are processed in parallel and, to a large extent, independently. At numerous lower hierarchic levels, the paths processing color-related and movement-related information are in principle separated, although both information-processing paths are concerned with orientation in space and stereoscopic vision. However, the degree of complexity of properties isolated in the image gradually grows. For instance, field V1 organizes particular image spots "received" from the retina into straight lines with different inclination angles. The famous experiment on the visual cortex of a cat (rewarded with the Noble Prize) showed that the strongest reaction of a part of neurons in field V1 is produced by a narrow strip of light (or darkness) or by a black-and-white edge (the remaining fields react to a circular spot of light/darkness). Each neuron prefers a characteristic orientation (angle) of a line or band that triggers the strongest reaction, while the neighboring neurons react to stimuli with similar characteristics/orientation. The reception fields of V2 cortex neurons (the next level in the hierarchy) are larger than in V1 cortex, and they are capable of more subtle reactions (e.g., they react to illusory contours).

At the following hierarchy levels, straight and curved lines are organized into more and more complex shapes that are integrated into various objects with a growing degree of complexity. There are numerous intermediate stages, still poorly examined, but one finally reaches to single neurons (in "higher-order" centers) that react selectively, for instance to human faces, while some of them are activated only by the faces seen in front (with a certain tolerance to deviation), others react only to the faces seen in profile.

Movement perception processing is carried out along a separate channel, parallel until a certain stage. At the following stages of integration, some neurons become interested in the direction and speed of object movement, while neighboring neurons react to similar speeds and directions. Other neurons are activated by approaching objects (that increase their angular size), while yet another group reacts to objects that are moving away. Color-related data are processed through a separate information-processing channel.

Thus, the process of perception is carried out as a set of at least partially parallel processes: the processing of data related to different visual aspects is carried out simultaneously, yet independently, in different brain areas. Finally, however, the highest degree of integration is reached (primarily in the temporal lobes, as it is supposed), i.e., different stimulus processing paths converge and particular visual aspects (shape, movement, color) are integrated (at least in a subjective sense) into a coherent whole. This is enriched with signals from other sense organs that are integrated into the final "sensory images." The temporal lobes and the neighboring fragments of the occipital lobes seem to contain (this is indicated by the consequences of brain damage) neural objects—formed and activated by receptor data—that correspond both to the more general concepts and categories (related to the semantic memory) and to particular objects and events (recorded in the episodic memory). It is also true that a considerable part of sensory functions of the higher order cannot be easily located within the brain.

I have already mentioned that the integrated sensory data from receptors are properly interpreted through a confrontation with memory records. At this point, the crucial question arises: is it at all justified to differentiate between higher sensory structures and the activation of corresponding memory records? In my opinion, it is not. For the process of perceiving cannot occur without understanding the perceived and the understanding of sensory stimuli is gauged against the semantic reference point, namely, the corresponding memory records (related to the events or concepts that correspond to what is being perceived at a given moment). The sensory images formed from current data simply "weave into" the already existing memory records (while participating in the formation of new memory traces). Therefore, it does not make sense to separate strictly the input data from the memory (and the fixed memory from the operating memory) or to differentiate between the hardware (the physical internal structure—the "wiring") and software, which can easily be done in the case of computers. The memory records in the brain are properly processed sensory data collected in the past that are used—as a reference framework—to understand and interpret new signals received from receptors.

Finally, integrated sensory images (some of them) reach the light of consciousness. Numerous data indicate that man is aware of only a small part of sensory images created in the brain. Lack of direct connection between consciousness and these images is confirmed, for instance, by the fact that the perception of familiar faces activates the "face region" in the brain of unconscious patients in a permanent comatose state. Consciousness, occasionally referred to as the highest receptor of the brain, operates (in my opinion) on the receptor-generated data at the highest stage of integration, although it also contains so apparently simple and elementary aspects of vision as the blue color or particular image points. I will argument further on that the understanding of both "blue-ness" and "point-ness" can occur only in the structural and functional context of the entire neural network and of the memory traces contained therein. This constitutes a part of the following problem, formulated long ago: why does consciousness receive some signals as sounds and others as images? Although, subjectively, they represent totally different qualities, they are founded on electric impulses that are identical (i.e., of the same type). This problem applies to the difference between the red and the blue as well. I believe this mystery can be explained by the activation of different fragments of the neural network by stimuli generated by different receptors (e.g., stimuli representing sounds and images or different colors), processed later on by different integration mechanisms. Under given circumstances, appropriate areas of the network receive appropriate nerve impulses and respond by sending back adequate impulses that amplify connections of these areas with the sensory paths leading to the areas as well as with other fragments of the neural network that are activated at the same time. In other words, the receptor signals are "correlated" in different cases with memory records that have been shaped in a different manner. Thus, the "meaning" or the "subjective quality" of different neurophysiological structures and processes is relationally determined through the reference to other structures and processes.

While the subconscious information processing in the brain, and the integration of various aspects of sensory data received from different receptors, seems to be carried out in principle in a parallel manner—as mentioned before—the majority of neurophysiologists agree that the brain processes that form the basis of consciousness are realized in a sequential, serial manner, i.e., one after another. So, how does it happen that the light of consciousness can simultaneously embrace so large a spectrum of phenomena, their various aspects and features, integrate such different qualities as shape, color, movement, and smell into one object, while containing thoughts, emotions, action planning and the self-awareness of its own existence? According to the available knowledge, the neurophysiological correlate of consciousness is not localized in a particular, strictly determined part of the brain.[4] It is rather a dispersed phenomenon (process) that involves neural circuits of substantial areas of the brain cortex (and, probably, of the thalamus—one of the oldest parts of the brain, in evolutionary terms). This often triggers the following question: how does it happen that all these different objects and aspects of consciousness merge, in our subjective reception, into a whole? In my opinion, the dispersion (of the neural correlate of) consciousness over a large brain area does not have anything to do with this, since the problem would not be any different if conscious processes could be located in a particular place with the volume of, say, three cubic millimeters. The sense of unity may be caused by intense communication between the brain areas functioning as "carriers" of different aspects of consciousness, irrespective of the distance at which the communication is carried out. Moreover, the conviction that consciousness is a unified whole is derived—in my opinion—from the subjective, introspective feeling/perception that does not have to correspond to anything "real." The most important thing, to my mind, consists in the fact that consciousness does emerge from the "non-psychic" activities of a network of nerve cells.

While the consciousness of (the processed view of) the external world is rooted in the sensory part of the cortex (the temporal and parietal cortex), the "consciousness of that consciousness," as well the consciousness of one's own "internal," psychological states (thoughts, decision making processes, action planning, emotions) and the consciousness of one's own "ego," i.e., simply the self-consciousness is located—in my opinion—in the prefrontal cortex that constitutes the decision-making center and the carrier of the operating memory of the brain. This is the place that receives (primarily, through the "Central Brain Bus") the integrated sensory images, and the place where they become subjects of consciousness, for the concept of consciousness becomes void and senseless if considered separately from the concept of self-consciousness. Because the operating memory is located in the prefrontal cortex, this is the place of the dynamic processes of associating and analyzing various data, of confronting the data with the records in the "fixed" memory. This is also the place of partly autonomous processes of long-range planning and thinking. It is here that decisions concerning current behavior are made (which triggers cascades of transcriptions than,

address where the ...

Generally speaking, the operating memory corresponds to (at least a part) of the short-term memory, including the fragment that constitutes (again, in man) the neurophysiological carrier of the content of self-consciousness, as it enters into the focus of consciousness (the phenomenon of attention participates in this process). Some of "memory traces" in the short-term memory, after appropriate transformation, may become (although they not always are) integrated into the long-term memory as relatively stable memory records. The transfer of memory traces (patterns) from the short-term to the long-term memory may be caused by multiple repetitions of such traces in the former, or by association of the traces with some important piece of information, or else by the general state of brain excitation.

Having begun from single receptor cell stimulation in sense organs by physical stimuli originating from the external world, we have already reached the highest level of integration and association of sensory signals received from different sense organs. The thought processes in the operating memory (that are conscious only partly and *post factum*) use the perceptual data being received and the stored memory records to make decisions concerning current and future actions. The above processes of extremely complex information processing constitute the highest hierarchical level of functioning of the human brain, which finds its reflection (at least partly) in the mind (as we remember, all this plethora of processing operations corresponds to an extremely elementary and strictly determined "decision" of a few nerve cells of a cnidarian that identify tactile receptor stimulation that is sufficiently strong to trigger the contraction of muscle fibers). This is where the very "core" of brain operation is located—at least from our anthropocentric point of view—since the "highest brain operations" (such as abstract thinking) that differentiate man from the rest of the animal kingdom occur here.

An expedition into the mountains does not consist only of climbing a summit, but it involves the return to the valleys as well. Let us begin the descent. The route is presented in figure 6.

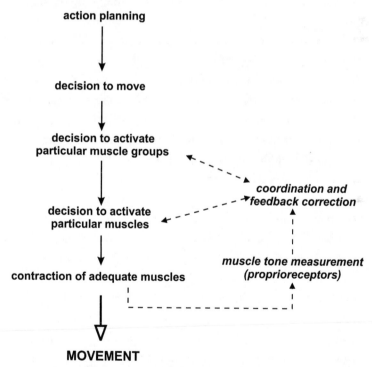

Fig. 6. Particular stages of the hierarchically organized transcription of general decisions into activities of particular effectors (muscles, in this case). This is an example of the descending (concretizing) path that leads from the more general to the more detailed directives that govern the stimulation of particular muscles.

The action planning in the operating memory yields a "tangible" effect in the form of a decision to behave in a certain particular manner, now or in the future, which most often is manifested by a movement of the entire body or its part (even such special activities as that of the enunciation of linguistic statements, writing or drawing are none other than properly coordinated forms of movement of certain body parts). At the moment of undertaking an action the decision to carry out a movement must be put into effect. As mentioned before, this is done gradually at numerous, hierarchically organized stages consisting in the transcription of general directives into more and more detailed instructions. The sequence and intensity of stimulation is determined for entire groups of muscles, single muscles, and finally single muscle fibers. The entire process proceeds in numerous places of the central nervous system. Its substantial part is carried out in the motor cortex, where particular areas correspond to the muscles of particular body parts, e.g., of the palm, calf, or trunk muscles, while the neighboring cortex areas "support" neighboring body parts, e.g., the fingers of the same palm, the brachial and antebrachial muscles of the same hand (a similar "map" of the body is located in the sensory cortex that collects tactile stimuli). Movement coordination, i.e., the determination of the temporal pattern of the (intensity of) contraction of particular muscles, especially in the case of complex learned skills, e.g., bicycle riding, is controlled by the cerebellum. It also participates in the feedback control of movements and body position. This is possible because muscles contain the so-called stretch receptors registering muscle states, which constitutes the "proprioception." The cerebellum receives data from these receptors, compares them with the "set values" of muscle stretching in a given situation, and goes on to modify them by increasing or decreasing the intensity and/or speed of particular muscle contraction. This is important not only for body movement, but also for maintaining appropriate body posture. Movements are also controlled by the prosencephalon that selects one of the available "locomotor programs." The brain stem initiates locomotor movements and controls their speed, while the spinal cord (another part of the central nervous system, located in the vertebral canal) covers the so-called cyclical patterns of movements, i.e., simple, inborn movement patterns that are responsible, for instance, for alternative stepping movement of limbs during locomotion. (It should be noted that although man uses only legs during locomotion, he simultaneously moves his hands and the movement is shifted by a half of the cycle with respect to the leg movement—this is a relic handed down from our four-legged ancestors). In general, many neural centers, complementary to a great extent, cooperate in controlling the entire locomotor system, while the following principle seems to hold true: the more complex, ontogenetically acquired, and conscious locomotor patterns are controlled by higher parts of the nervous system (especially the brain cortex), while the behavioral patterns that are inborn, simpler, unconscious, and older in evolutionary terms (see the discussion below of the unconditioned reflex of retracting the hand that has touched a hot object) are controlled by lower parts of the central nervous system, including the spinal cord.

The stage of activating the relevant effectors (muscles) terminates the entire chain of transmitting and processing information through the "infrastructure" consisting of the nervous system, receptors, and effectors. Its evolutionary role is clear: it is to ensure a behavioral reaction of a given organism to environmental stimuli that is sufficiently optimized to maximize its chances of survival and leaving as great a number as possible of as efficient offspring as possible. The brain is the highest element of this chain, but it is also nothing but an element in the chain. Many simple or even more complex animals can do without it. On the other hand, it is the brain—in the most advanced stage of its development hitherto known—i.e., the human brain—that made it possible for the third level of reality to emerge (above the physical and biological levels), namely, the level of the psyche/mind (the subjective sphere of mental experiences).

4.

THE GENERAL STRUCTURE
OF THE NEURAL NETWORK

All the above-discussed brain functions are realized by the neural network consisting of nerve cells (neurons). The properties of the network are partly derivatives of the properties of single neurons (discussed in chapter 2), while they partly emerge as a result of the specific (structural and functional) organization of the entire network. There are three essential aspects of this organization. First, it is crucial which neurons send signals (excitatory or inhibitory, depending on the type of the sender) to which neurons in a given network. (It is interesting that some neurons send signals to themselves, which can lead to spontaneous generation of signals without external stimulation—this is called self-stimulation—or to self-moderation of the activity of a neuron, i.e., to self-inhibition.) As mentioned before, since the number of neurons in the brain is huge (about a hundred milliard) and so is the number of synaptic connections of each neuron (several thousand), only a very small (almost negligible) fraction of all possible connections can be realized and it is the pattern of existing connections that defines the specific characteristics of a given brain as the data processing mechanism that transforms environmental stimuli into the behavior of the organism (of man or an animal). Second, it is important how many synaptic connections there are between particular neurons (of course, it is also important, whether they are excitatory or inhibitory in character). Finally, the information processing carried out by the neural network is conditioned by the excitation threshold (the connection weight) of particular synapses, i.e., the degree of "patency" of neural routes established by the synapses. All the above-mentioned aspects of the global organization of a neural network—i.e., the existence or lack of structural and functional connection between different neurons, the number and weight of synaptic connections—can change in the process of learning. Thus, the structure of the entire network is capable of evolving and gathering experience, as the neural network of the human brain does in the course of ontogenesis. It is important for further considerations that all the discussed processes of sensory data integration, association of the resultant sensory images with other images and memory records, and of planning future actions and transcribing decisions into movements occur in an organized—as described above—neural network consisting of nerve cells, and the very process of creating memory records is related to the evolution—described above— of the structure of this network under the influence of successive stimulations of its inputs (and of autonomous processes of information analysis and processing, i.e., thinking). Thus, the lacuna between the functioning of a single neuron—as described in chapter 2—and the processing of information realized in the macroscopic scale by the entire brain—presented in chapter 3—is in a way filled in. This chapter will focus precisely on this issue (i.e., on the structure and functioning of the neural network at the intermediate level between a single neuron and the brain).

The entire nervous system of an animal or man is nothing but an appropriately organized (structurally and, above all, functionally) network of nerve cells. It is an autonomous, and, to a great extent, closed system of circulation of nerve impulses (because stimulations of receptors and activities of effectors do not constitute impulses of the type), where the outputs (axons) of some neurons are connected (through synapses) to inputs (dendrites) of other neurons. In a neural network, all neurons receive signals from (and send signals to) other neurons. Some nerve cells are able to generate signals "by themselves," without being stimulated by other nerve cells. However, since stimulations (or inhibitions) are as a rule transferred from some neurons to other neurons in a neural network, such network contains neural "routes," "circuits," and "structures," i.e., sequences and highly ramified series of neurons that carry and process pieces of information. The brain also contains the so-called reverberating circuits (consisting of numerous neurons), i.e., closed and cyclical routes of nerve signal transmission, where excitation stimuli can circle around without external stimulation[1] (they should not be mistaken for the above-mentioned self-stimulating neurons). All neural routes and reverberating circuits are, of course, mutually interconnected (directly or indirectly, structurally or functionally) into one, highly integrated SYSTEM (although it contains partly independent modules). It may be the case that the processes that become (or at least can potentially become) the content of the light of consciousness are responsible for the integration of the variety of brain subsystems at the highest level. I believe that the above-mentioned system constitutes a certain objectively existing entity, a certain new quality that is a derivative of the specific, goal-oriented complexity of the SYSTEM, as well as of certain intentional information related to the complexity.

This network of mutually interconnected nerve cells—autonomous to a great extent and constituting a certain qualitatively separate "world" (in the sense that is very important from the psychological point of view)—breaks off in the blind alleys of its inputs and outputs, i.e., receptors and effectors. It breaks off in the sense that the information coming into the receptor cells from the environment and the information "sent into the world" by the effector cells is no longer in the form of neural electric impulses. The receptor cells, for instance, may be viewed as "half-neurons" with the signal-receiving side compatible with a certain type of physical stimuli from the external world, which the cell is tuned to, while the signal-transmitting side generates signals in the form recognizable for the nervous system, i.e., as the action potential and neurotransmitter release into the synaptic cleft. The effector cells (e.g., muscles), in turn, are "half-neurons" that are reversed with respect to receptors. They receive signals in the form typical for the neural network (a neurotransmitter that generates an increase of the potential on the postsynaptic membrane) but transform them into an action that assumes a totally different, physical form (e.g., a muscle contraction that generates movement or secretion of chemical substances by glands). "Neutral" physical phenomena, such as electromagnetic radiation (registered by the sense of vision) or mechanical work (resulting from muscle operation) assume—in the context of an animal (or human) nervous system—a certain informative value, useful in the realization of a given goal. The superior, purely biological goal—established in the process of evolution—consists of course in surviving and producing a possibly great number of offspring.

Thus, the information processed by the nervous system has

a biological value. Everything that is outside the SYSTEM of the network of neural connections (including the material substance of the network) assumes, in a sense, a totally different quality than the formal (abstracted from its material base) pattern of the network operation (information processing). Thus, the material carrier of mental processes, belonging to the physical level of reality, should be clearly differentiated from the degree—and type—of the complexity of the entire information processing system (that can be described in terms of cybernetics and the theory of information) that belongs, in my opinion, to the biological (neurophysiological) level of reality (and in certain conditions—to be discussed further on—it can form the basis for the mental level of reality). I believe that both levels based on complexity—i.e., the biological and the mental levels —exist in a no less objective manner than the physical level.

If one considers the tremendous structural diversity of neurons (different cell body shape and varying length and branching of dendrites and axons), the number of (both excitatory and inhibitory) synaptic connections with other neurons variously arranged in space, the huge number of combinations of excitation thresholds of particular synapses, as well as the variable sensitivity of impulse-generating axon root of particular neurons, it seems reasonable to assume that there are no two identical nerve cells in the human brain. Therefore, each of the cells realizes a slightly different "logical function," i.e., it has a characteristic (unique) pattern of assigning various output signals to a huge number of potential combinations of input signals (the pattern can, of course, evolve in the process of learning). Does this determine completely, or at least primarily, the role of a given neuron in the neural network? Of course not! At least equal importance should be ascribed to the place of a given neuron in the network, to its function in a greater entity, as well as to the neurons it communicates with and the manner it communicates with them. All these factors are necessary to determine the full "significance" (meaning) of a given nerve cell within the entire neural network. The "significance" of the other neurons is, in turn, determined by the functional connections with still other neurons, and so on. Finally, we arrive at the conclusion that the "sense" of each nerve cell is determined by the context of the entire neural network, all the neurons, receptors, and effectors comprised by the network, i.e., by the abovementioned SYSTEM. Therefore, neurons cannot define "by themselves" their own role within the totality of brain processes. It becomes possible only in relation to the entire, inexhaustible plethora of other neurons. Now it becomes clear why some signals from receptors appear as auditory sensations in our consciousness, while others take the form of visual sensations, in spite of the fact that the signals can be identical in physical terms (as series of identical impulses of a determined frequency), as the brain "treats" them as signals of a given type, because they reach the relevant brain centers, where they are adequately processed (it should be clear that auditory and visual stimuli are integrated in a radically different manner). This is the base for the subjective sense of totally different qualities. The difference between red and blue colors has a similar basis, although both appear in our consciousness as visual qualities, due to greater similarity between the respective mechanisms of integration and association.[2] These are, of course, very simple examples. However, even the most complex features and meanings are identified in the brain in principle by the same method, i.e., by relating them to other features and meanings. This relative and relational significance of elements of a system, established by their relation to other elements, is referred to as connotative significance (as opposed to the denotative significance, i.e., direct assignment of a significance of an element to its referent, e.g., a subjective perception of the red color—the electromagnetic wave of a determined length; linguistic elements signify denotatively, as the word "dog" refers directly to a dog). As it will be demonstrated further on, this feature of a neural network is of great importance for the essence and structure of the conceptual network in our mind and for the worldview created within it. This, in turn, conditions—in an unimaginably significant manner—the scope of our cognitive abilities and the nature of our cognition.

Of course, the neural network is not a uniform formation. Its different domains are interconnected with other domains to a greater or lesser extent, and they show varying degrees of internal autonomy. We are talking here not about purely physical localization within the brain, but about the number of connections and functional integration. Different domains may be spatially "dispersed" over a large area, e.g., of the brain cortex, and their networks may overlap. (It is not necessary for neurons whose cell bodies are close to each other to communicate with each other in a more intensive manner than with nerve cells located far away; on the contrary, the length of processes—axons in particular—ensures effective transmission of signals over large distances, while the neighboring neurons may know nothing about each other). Of course, the functional brain "centers" of the type are often hierarchically organized: the centers responsible for more general functions contain subcenters that realize more specific tasks, while these consist of even more specialized sub-subcenters. It should be clearly underscored, however, that it is often difficult to determine clear borders of a neural network corresponding to a given brain center (as opposed to a computer, where it is easy to distinguish particular subassemblies). This applies also (or perhaps: primarily) to memory records, especially the records of the semantic memory concerning particular objects, categories, concepts, and names. This fact—as will be discussed further on—is directly related to the fuzziness and ambiguity of significance, the lack of absolute precision of the concepts we operate with. The connotative character of the functions of neurons within the neural network is finally responsible for these features.

An important question arises here: to what extent is the structure of the neural network and the tiniest details of its organization (i.e., the properties of each synaptic connection) inborn, i.e., conditioned by the genetic record of the organism, and to what extent is it shaped during ontogenesis, through acquired experience, gathered memory records, and as a result of incidental events. In lower animal organisms, the in-born component is dominant or it totally determines the structure of connections between neurons, although elementary learning processes were observed even in animals with a relatively low taxonomic position, e.g., in snails, which means that the pattern of synaptic connections in the animals is modified through experience acquired in the past (thus, the behavioral patterns of such organisms' reactions to external stimuli are also modified). In man, both components play a very important role. The in-born component (the result of the accumulation of "learning" and incidents during biological evolution) conditions the general brain structure, the location of cell bodies of neurons, the routes taken by nerve fibers, as well as the predisposition to create and modify synaptic connections. The basic mechanisms of integrating signals from receptors are also inborn, for instance the mechanisms that are responsible for making some perceptions visual, and others auditory. Broadly speaking, we come to this world with an immanent predilection for integrating various elements of sensory images into spatial, temporal, causal, and other relations. Another important thing: our brain has inborn, strictly localized Broca's and Wernicke's areas that allow man (thanks to their general genetically conditioned functional architecture) to learn language, and when it is learned, they are responsible for the enunciation of sentences and ordering of words into grammatical sequences ("expressing thoughts in words") and for understanding speech, respectively (it should be remembered, however, that these centers must acquire the details of their functioning during the ontogenesis; for instance, they must be trained to use a particular ethnic language). Thus, our brain "knows" a lot about the world and its own functioning (it has been "trained" by biological evolution) before we are even born.

However, a huge part of the structure of the neural network in the human brain is acquired during ontogenesis. This applies in particular to the pattern of local (in the functional sense) connections between neurons, especially in the regions of nerve cell networks that are responsible for complex cognitive and decision-making activities. And thus, ontogenesis shapes the aspects (fragments) of the neural network that are related to the high-level integration of sensory data, association (of sensory images, memories, single events, features, rules, event coincidence), episodic memory (the memory of particular events and facts), semantic memory (understanding detailed and general concepts, categories, rules, dependencies, language skills), procedural memory (coordination of movements, acquired locomotor patterns, including language skills), and so on. It is during ontogenesis that we acquire a detailed (constantly evolving) pattern of synaptic connections (including their weight) related to higher cognitive and decision-making activities, such as recognition of complex objects, identification of complex rules that govern the behavior of sets of objects, establishment of general categories, thought processes (problem solving), planning and decision making, and, finally, implementation of complex loco-motor directives. During ontogenesis, we also learn to under-stand and use language, moral principles, arbitrary cultural values, and the principles of logic. We also develop (to some extent) the aesthetic judgment and acquire the knowledge that has been gathered by the society.

I will try to present in general terms—on the basis of the above-described "learning" potential of a single neuron—how the acquired structures in the neural network (or at least some of them) are established at the neurophysiological level. I will limit the discussion primarily to the integration of sensory stimuli, although I believe that similar principles apply to the decision making and locomotor parts of the brain cortex. I will begin with the inborn "basis" of the functional neural structures that are acquired during ontogenesis.

Let us go back to the multistage, hierarchically organized integration of signals received from receptors, as presented in figure 5 (see page 41). At the input side, quanta of electromag-netic radiation stimulate the light-sensitive cells in the retina of an eye, i.e., particular image pixels (by analogy to picture ele-ments on a monitor screen). How does it happen that the distri-bution of excitations over the retinal surface is forced by the brain into the corset of certain categories and interpreted as spots, contours, and lines at different angles, as the entire color scale with light and shade effects, as objects, time, space, move-ment, causal relations, etc.? After all, no such objects as lines or individual objects exist objectively in this image. It is the work of our eye and brain that particular pixels are grouped into com-plexes and isolated as autonomous entities, phenomena, or properties. But, why does the brain "prefer" to pick up lines and not some chaotic distribution of spots or the fact that something is simultaneously present in the top right and bottom left corners of the visual field? This preference is imposed by certain patterns of connections between neurons within the neural network. I will refer to such patterns as the integrative structures. They are hierarchically organized and they are responsible for successive stages of sensory data integration and for the creation of complex sensory images, as shown in figure 5. There is no mystery in it. I am going to give several examples of the operation of such structures to demonstrate that no black magic is needed to explain these processes. Here is the most trivial example: pixels (stimulated light-sensitive cells) are integrated into a spot, when a neuron collects signals from several (dozen) rods and/or cones that have been simultaneously activated. Violet color can be produced as a result of the same excitation intensity of a neuron by "red" and "blue" cones (such a neuron has to be "connected" to such cones and it has to sum up and compare the relative intensity of excitation of both types of cones). Objects are produced as a result of grouping similar pixels or spots in one place on the image, while object movement is extracted from the fact that object images in successive temporal units occupy neighboring spatial positions on the retina, which reflects the constant displacement of an object in the visual field. Lines are perceived when a stimulated light-sensitive cell has stimulated neighbors in a direction, while the neighboring cells in the perpendicular direction are not stimulated. Shapes are produced at a higher level of integration by assembling lines, spots, etc. In each of the above-mentioned cases, there is of course a neuron or a group of neurons properly (in the functional sense) connected to each other and (directly or indirectly) to receptor cells. I am convinced that an average software developer would not find it difficult to write an algorithm that would be capable of imitating all the above-mentioned simple examples of visual data integration if he had the possibility of accounting for formal parameters of excitatory and inhibitory synapses and for different weights of synaptic connections (I think I could do it myself if I made a little bit more effort to do so). Similar general rules of integration apply, of course, to other senses as well, although the details of integration are certainly different. For instance, sounds are not organized into lines, shapes, and spatial images (at least in man). It seems probable, however, that bats create a spatial image of the world on the basis of auditory data that is much more similar to our visual than to the auditory image. This shows that the manner of sensory signal integration is more important than the type of these signals (i.e., which receptor cells send the signals and what physical stimuli they are sensitive to).

I provided examples of the operation of very simple integrative structures responsible for picking relatively elementary features and aspects from the distribution of pixels over the image surface. They constitute the lowest hierarchical level. Integrative structures of a higher order compose elements "produced" by integrative structures of lower order into more complex objects that are, in turn, integrated into even more complex entities (e.g., human faces) by the integrative structures of an even higher order. Figure 7 presents the schematic operation of integrative structures at successive levels of the hierarchy. For the sake of simplicity, only two hierarchic levels are presented, although the actual hierarchy may consist of numerous levels. The figure shows numerous signals (1, 2, 3 . . .) from particular receptor cells that are integrated into a simple feature A by the integrative structure of a lower level. This and other features (B, C, . . .) extracted by other low-level integrative structures are integrated into a complex feature represented by the Roman numeral I by an integrative structure of a higher order. As I have said, the actual number of integration stages is much higher.

Hierarchical integrative structures

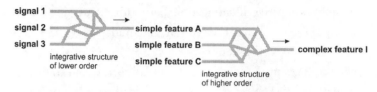

Fig. 7. Schematic diagram of the hierarchically organized functional integrative structures in a neural network. They "organize" signals from single receptor cells, first into simple and then into more and more complex properties.

The integrative structures of the lower orders are (in my opinion) innate in man. It means we have a genetically coded ability to perceive spots, lines, colors, and movement. On the other hand, the integrative structures of higher orders are developed during ontogenesis to reach their final form. It is in the process of developing and modifying the system of connections within the neural network that we learn to perceive, identify, and understand (there is no perception without understanding) more complex sensory images. An x-ray image of the chest is my favorite example. As a layman, I can distinguish only dim outline of ribs and lungs in such image, while an experienced doctor is able to pick out a whole plethora of details that help him identify numerous lesions. Where I see nothing but chaos of obscure spots, his highly developed integrative structures automatically organize the seemingly senseless distribution of spots into a pattern that carries much more information.

Here is another example illustrating the process. I used to ring bats. I caught them in nets or looked for them on high cave ceilings in winter. After several years of practice, I was astonished to discover that I can identify bats faster and with greater ease, in spite of my rather poor eyesight, than those who have good eyesight, but are novices in chiropterology. By trial and error, after finding hundreds and thousands of bats, i.e., after numerous positive identifications and failures to identify correctly these creatures, the relevant integrative structures in my brain learned which seemingly chaotic arrangements of spots on cave ceilings in the glow of a flashlight correspond to bats and which do not. The satisfaction of correct identifications intensified the neural connections that produced them, while the disappointment of misidentifications weakened the connections responsible for the errors. (In the next chapter, I will discuss the manner in which the reward-punishment system and the corresponding emotions lead to the strengthening of "advantageous" and to the weakening of the "disadvantageous" connections in the brain.) I think that the reader can easily find similar examples of the evolution of integrative structures in his own life.

Certain higher-level structures of connections within the neural network that are developed totally during ontogenesis will be termed associative structures. What are they? I will begin by indicating that it is impossible to define a clear-cut distinction between associative structures and integrative structures of higher orders, especially those that are not innate but acquired during ontogenesis. In the case of associative structures, however, their genesis and essential characteristics are dominated by associative processes that occur both as a result of temporal and spatial coincidence of sensory images, and as a result of thought processes (e.g., association by analogy). They may also, at least potentially, operate on the data originating from many or even all sense organs. Associative structures are strongly related to memory records, especially those concerning semantic memory and episodic memory. In my opinion, these structures—their activation and modification by sensations and thought processes that lead to the establishment of new associative structures—constitute the neurophysiological "carrier" of mental objects and mental processes. Some of them become the substance of our consciousness. In general terms, they correspond to broadly understood concepts—according to my definition of the term—at the mental level (see chapter 6 on the nature of mental objects).

I decided to divide associative structures (arbitrarily to a large extent) into structures of lower and higher order. The establishment of associative structures of the lower order is schematically presented in figure 8. Repeated temporal coincidence of neuron excitations and activations of certain nerve routes responsible for various complex features "extracted" as a result of processing the signals from receptors by integrative structures leads to "opening" the neural interconnections

Establishment of associative structures of lower order

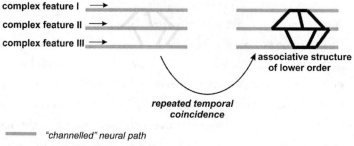

complex feature I

complex feature II

complex feature III

repeated temporal coincidence

associative structure of lower order

"channelled" neural path

"un-channelled" neural path

object in neural network (highly "channelled" neural circuit)

Fig. 8. Schematic diagram demonstrating the formation of associative structures of the lower order. Repeated temporal coincidence of stimulation (especially by signals from receptors) of integrative structures corresponding to a certain set of complex properties leads to the establishment of an associative structure of the lower order, i.e., a certain "object" in the neural network, within which synaptic connections have a low excitation threshold, which is why it is easy to activate this object as a whole, even with a relatively weak signal.

between the routes that were formerly able to function independently to a considerable extent. If a neural connection is not "patent," it may mean that the number of synaptic connections is small (or even there are none) and/or that the excitation threshold of synapses is high. Opening functional passages between different nerve routes in a neural network would consist in creating new synapses and/or increasing the weight (reducing the excitation threshold) of the existing synaptic connections. I have already described the process, while discussing the characteristics of a single neuron, especially its ability to learn. Highly patent connections that "associate" various complex features (aspects) into one whole constitute a certain object within the neural network: the associative structure of a lower order that corresponds (or can correspond) to an object (thing), fact, or aspect of reality. If such structure is established, it is not necessary for sensory stimuli to activate simultaneously all the complex features constituting a given object in the neural network in order to activate the corresponding associative structure. On the contrary, a weak stimulation of some neural inputs of a given structure leads to the activation and "replay" of the entire formation (numerous nerve signals of high intensity run through the neurons constituting the structure) together with the features, aspects and fragments that are not directly activated at the moment by sensory data. For instance, we reconstruct the entire animal from a fragmentary image of a cat running through high grass or from the neighing of a horse that is located outside our field of vision. We do not even need sensory stimulation to evoke an indistinct image of a cat or horse. We do this in our thoughts, in imagination, in our dreams and hallucinations, and in the process of understanding the concepts of a "cat" and a "horse."

While associative structures of lower order are related to

rather concrete, tangible objects, events, features, and aspects of the external world, the associative structures of higher order refer to general categories, laws, rules, and relations. The latter are created not only as a result of temporal, contextual, and situational coincidence of various associative structures of lower order, but primarily as a result of thought processes—autonomous to a large extent—that associate various neural "objects" according to their own logic conditioned either genetically or by experience acquired during ontogenesis. The associative structures of higher order can be created, for instance, through picking out repeating patterns in associative structures of lower order. Therefore, their form is to a lesser extent a derivative of signals from receptors processed by integrative structures, and, to a greater extent, a derivative of built-in (or learned) neurophysiological mechanisms that are to a large extent arbitrary (this fact is relevant for the discussion of the correspondence of our world vision to the real world in its immanence, to be presented further on).

The establishment of associative structures of higher order is schematically presented in figure 9. It can be seen that the process is similar, in general terms, to the establishment of associative structures of lower order, as opening neural passages (this time, between various associative structures of lower order) also leads to the creation of a certain object within the neural network, and the object is, in a sense, both a sum, a derivative and a resultant of its components. The main difference consists in the fact that the process occurs at a higher level of the hierarchy of complexity and of abstraction, much further off the "directness" and tangibility of sensory perceptions. Automatic integration and association is replaced here by active and self-generated association. The associative structures of higher order have more in common with the semantic memory (associative

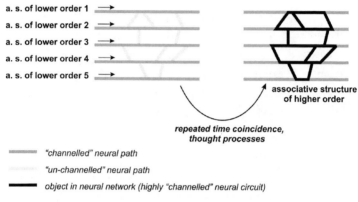

Fig. 9. Schematic diagram demonstrating the formation of associative structures of the higher order. Repeated temporal and situational coincidence of stimulation (e.g., due to thought processes) of a set of associative structures of the lower order leads to their association and to the establishment of an associative structure of the higher order that constitutes an autonomous neural "object" at the highest level of the hierarchy.

structures of lower order have a stronger relation to the episodic memory) and they provide a reference framework for the processes of planning and decision making that occur in the prefrontal cortex, where the operating memory is located. Indeed, I would venture to claim that thought processes are nothing but the activation of the already existing associative structures of higher order, their modification and the establishment of new associative structures of higher order.

Figure 10 summarizes briefly everything that has been said above about the integration and association of sensory data by the neural network in the human brain. Signals from receptors, devoid at first of any subjective significance, are interpreted and integrated in a multistage process by hierarchically organized integrative structures that pick out from the current pattern of receptor cell activation these features and aspects that, during

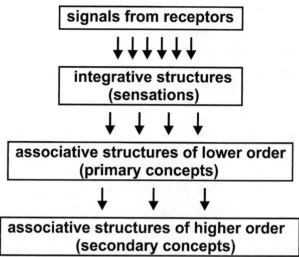

Fig. 10. Schematic presentation of a sequential organization ("interpretation") of signals from receptors by integrative structures and associative structures of the lower and the higher order. It shows mental objects corresponding to these structures, namely: sensations, primary concepts, and secondary concepts. The figure is to demonstrate that sensory stimuli may not only shape structures within the neural network, but that they can be interpreted by these structures.

our evolution, turned out to be important for the survival and reproductive success of our ancestors. This is how sensations (understood as subjective mental objects) are formed—which will be discussed later on—although I believe that associative structures are indispensable for understanding sensations. Associative structures of lower order are established by the association of complex features and aspects coinciding in time and space that form sensory images created by integrative structures. In the mental sphere, they correspond to primary concepts (detailed discussion will follow) concerning concrete, individual, and tangible objects, facts, features, and aspects of the world. On the basis of associative structures of lower-order, autonomous thought processes characterized by self-generated and spontaneous dynamism establish associative structures of higher order that have found their subjective correlate in secondary concepts referring to general and abstract entities, categories, laws, and dependencies. Thus, by following the main idea of this book, we are halfway to bridging the gap between neurophysiological processes and functional structures, on the one hand, and subjective mental phenomena, on the other hand. Chapter 6, focused on the nature of mental objects, will take up our construction work from the other side, i.e., from these objects. In chapter 7, where the essence of self-consciousness is discussed, both parts of the bridge will be connected with the keystone. I will attempt to show there how a certain type of complexity of neural network functioning could lead to the emergence of the subjective sphere of our sensations, i.e., to the emergence of the third level of reality—the mental level (following the physical and biological levels).

Generally speaking, both integrative and associative structures form the *cognitive apparatus* of the human brain, within which a certain (instrumental and operating) view, projection, representation of (certain aspects of) the external world is formed. The role of the named cognitive apparatus consists, on the one hand, in taking advantage of the already possessed worldview to interpret and understand complex combinations of sensory data constantly flowing into receptors and, on the other hand, in using the data to develop and modify the existing worldview. All this is finally aimed at planning future actions, solving various problems (either encountered in confrontation with the environment or posed by an individual to himself), predicting events in the environment etc., i.e., at realizing broadly understood thought processes. Our science and philosophy developed as an extension and sublimation of these functions that developed biologically during evolution.

The goal of possessing and using the worldview—together with the knowledge of objects contained therein and of its governing rules and its regularities—seems clear. To take decisions, plan future activities, and solve (current and expected) problems, one has to know how to account for and anticipate the features of the objective reality and the spatial, temporal, and causal relations therein. It is necessary, for instance, to predict what can be found behind a particular hillock, how an encountered predator or prey is going to behave, what benefits and hazards can be expected from the coming rain or the oncoming winter, which sexual partner will ensure the highest vitality of the offspring, and so on. To achieve this, the cognitive apparatus must be equipped with a worldview (established during the biological evolution and/or ontogenesis) that reflects just these aspects of the world that are crucial from the point of view of survival and reproductive success (dissemination of an individual's genes) and it must use the view at every moment. The "worldview" at the disposal of a few nerve cells of a cnidarian is not sufficient for so highly organized a being as man, as it is capable of recognizing only one aspect of the world (the presence of something big enough to produce mechanical stimulation) and its "repertoire of responses" consists of only one reaction (body contraction). It is the tremendous complexity of the neural network in the human brain that allows us to take into account the unbelievable variety of aspects of reality and react to them appropriately.

Below, I will attempt to trace briefly the evolution of neural networks (nerve systems) and their corresponding worldviews from a few neurons of the cnidarian to the human brain equipped with self-consciousness. Before doing so, however, I have to indicate clearly that neural networks have no monopoly on reflecting certain features (aspects) of the external world within the structure and functions of biological organisms, in order to adequately react to these aspects. For, each goal-oriented property of living organisms (formed by the biological evolution to maximize the chances of survival and reproduction) must account for the facts of the objective reality (its purely physical aspects as well as other organisms), including the fact that the substance of such organisms is of biological nature. This applies primarily to various networks of regulatory mechanisms at different hierarchical levels of the functioning of biological systems: for instance, the biochemical, genetic, or hormonal networks. To substantiate the above, I will give a few examples. The concentration of various amino acids, indispensable for the synthesis of proteins and other metabolic processes, may change in the external environment of a given cell. The demand for particular amino acids is not constant in time either, as it depends on the current metabolic status. Thus, it is necessary for a cell to start synthesizing amino acids from other organic compounds when their external supply does not meet the current needs, and to stop the synthesis when the availability of a given amino acid is relatively high. The metabolic network (consisting of metabolic routes and cycles) in a cell is equipped with appropriate regulatory mechanisms that allow the cell to cope with this problem. The first enzyme on the metabolic route leading to the production of a given amino acid is inhibited by high concentration of this amino acid, while its low concentration unblocks the enzyme and leads to its increased activity. The mechanism of negative feedback allows the concentration of the amino acid to be maintained at a set level and the demand for the substance to be met, irrespective of the variations in the amino acid availability in the environment. In the network of enzymatic reactions constituting cell metabolism one can find much more of such "intelligent" regulatory mechanisms that compensate for potential variations in the external conditions and for disturbances that originate both in the environment and in the organism itself (in its current state and the correlated needs).

An impressive number of such mechanisms can also be found at the genetic level, where they control the processes of reading and interpreting genetic information (they can be reduced to a temporal and spatial pattern of the intensity of production of various proteins in particular cells and tissues). The most famous example is found in the operation of the so-called *lac* operon in bacteria (there are many other operons, but the lac operon was the first to be discovered). In case the environment lacks lactose, the protein called repressor inhibits the synthesis of three enzymes that are responsible for the decomposition of lactose, since the production of enzymes that have nothing to do at the moment would be a waste of energy and material. When lactose appears in the environment, but glucose is lacking (another, much more common sugar that is easier to assimilate) a molecule of lactose binds with the repressor and deactivates it, which leads to unblocking of the synthesis of the enzymes that belong to the lactose metabolism route (transcription of the coding fragment of the operon to the mRNA section that contains copies of genes of these enzymes, followed by the translation of the sequence of the nucleotides in mRNA to the sequence of amino acids in appropriate proteins). Since an alternative (to glucose, for instance) source of energy (and building material) is available, a bacteria cell can continue to grow and produce daughter cells (by cell division). This regulatory mechanism also has its clear and precise goal realized by the following sequence of events: stimulus (presence of lactose)—perception (binding with repressor)—processing (unblocking the lac operon at the place of repressor binding)—reaction (synthesis of appropriate enzymes and lactose decomposition). The sequence is formally analogous to the sequence of information processing realized by a nervous system (especially by its simplest forms, such as a few neurons of the cnidarian).

Adequate networks of regulatory mechanisms—that have to constitute an operative, instrumental reflection of the features of the physical reality—can also be found at the physiological level. The network of hormonal regulation is a good example. Depending on the situation—whether we have just eaten an abundant meal or have been starving for days (which depends, at least in the case of animals, on the current availability of food in the environment)—the level of glucose in our blood is either excessive and has to be reduced to avoid diabetic coma, or insufficient and needs to be raised—e.g., by mobilizing the reserves of glycogen—to meet the energetic demand of various organs. The task of maintaining the level of glucose in blood within the set range is realized by a pair of antagonistic hormones, namely: insulin and glucagon—the first decreases sugar concentration, while the second increases it.

A certain "reflection" of the properties of the external world does not need to be sought in regulatory mechanisms built into the networks of biochemical, genetic, or physiological processes. For instance, the very shape of the catalytic cavity of the enzyme constitutes a representation of the spatial structure of the substrate. *De facto*, in a certain, very important (functional) sense, the active center of the enzyme (shaped during biological evolution) still "knows" more about the spatial distribution of electron density in the substrate particle than we do, having at our disposal the entire mathematics, quantum physics and chemistry. Summing up the course of thought, one can say that (in evolutionary terms) practically every goal-oriented aspect of the structure and functions of living organisms constitutes, to some extent, a reflection of certain features of the external world. Neural networks do not constitute an exception in this context.

On the other hand, it should be remembered that neural networks are the only ones to create a view of the external world, characterized by so tremendous wealth and complexity, available to us by introspection. It is only here that the structure of the view fits qualitatively to our subjective experience (it is difficult to imagine a psychic experience where the biochemical or hormonal network would constitute the subject matter). Therefore, while tracing the evolution of the worldview in living organisms, in the context of the emergence of the consciousness of its own existence gained by the worldview (i.e., the projection of the view onto itself, which will be discussed in chapter 7), I will limit the discussion to neural networks.

Let us recapitulate: the view of the external world, created in the course of biological evolution and ontogenesis, is nothing but an instrumentally efficient operating representation of objects, features, and laws of this world that are important from the point of view of the survival of an organism and of the dissemination of its genes. The network of nerve cell connections "entwines," in a sense, the facts (aspects) of this world just as a spider web may entwine the surface of a sculpture (I will exploit this meaning-rich metaphor later on). What can evolve is the diversity and resolution of various functional subassemblies of this network that determine the number of the aspects of the world that can be projected and the accuracy (adequacy) of the projection. The levels of both diversity and resolution have grown, as the very nervous system and the systems of receptors and effectors have been getting more complex. We will see how they have changed in the course of biological evolution of the world of metazoa.

Let us begin with our cnidarian equipped with a "nervous

system" consisting of a few cells. As a result of receptor stimulation, the nerve cells cause contraction of muscle-like cells
(epitheliomuscular cells) along the anemone's body axis that
results in the contraction of the entire body. Does the "neural
network" of this sea anemone constitute a view of the external
world? We understand perfectly well what it means that our
mind (and consequently, our brain) contains an image of a material object existing in the external world, e.g., a red rose, or a
reflection of a property, e.g., the fact that fire scalds. Does the
receptor-neuron-effector system of a sea anemone constitute a
representation of a feature or aspect of the objective reality? Of
course it does! Its operation has the following significance: "if
something stimulates you mechanically, it is probable that the
something is a predator, so that it is better to retract the body to
avoid being eaten." We are dealing here with an inscription
(extremely simple, to be sure) of an object from the external
world—a predator and the related threats—into the functional
"structure" of the "nervous system" of the sea anemone. The
anemone could react in a different manner to a mechanical
stimulus, e.g., it could release gamete or even stretch toward the
predator! It does not do this, because such a reaction would be
biologically harmful, as it would lead to the elimination of the
sea anemone and its genes. Thus, the sea anemone "knows" in
some way what could be expected from a predator and it presents an adequate behavioral reaction. It can be concluded that
the "neural network" of the anemone contains a residual form of
the concept of a "predator." I will provide arguments that all
concepts in our brain, even the most complex and abstract ones
(representing either a real aspect of the world, or imaginary
objects, such as angels or centaurs), are derived from extremely
primitive nuclei of concepts, like those presented above. All the
concepts are of instrumental character, and they are finally

aimed at adequate transformation of sensory data into the animal's behavior.

The functional significance of a few nerve cells of the cnidarian is of course strictly determined (including the excitation threshold) by the "semantic context" of those cells, i.e., by the type of receptors and effectors they are connected to. In neural networks consisting of numerous neurons, receptors and effectors, all the neurons, receptors, and effectors constitute the adequate context defining the significance of a given nerve cell (the view of the world is, of course, developed in a bit more detailed manner in such networks). This is the above-discussed functional connotative significance of neurons, i.e., the operational reference of neurons to other neurons as well as to receptors and effectors, which finally determine the goal and sense of the entire nervous system.

The degree of complexity of the receptors-neural network-effectors system (the carrier of the worldview) rose considerably during the biological evolution from the stage of a cnidarian equipped with a few neuron-like cells. The majority of the cnidarians (the group of the most primitive tissue animals that already have nerve cells, where sea anemones belong) is equipped with a loose network consisting of several neuron-like cells distributed in an almost uniform manner all over the body wall and provided with equal priorities. I have already indicated that they can conduct stimuli in all directions, as opposed to definitive neurons. In spite of their low taxonomic position and primitive structure, coelenterates are equipped with quite an extensive set of receptor cells sensitive to mechanical stimuli (touch), chemical stimuli (smell), and visual stimuli (sight). They have also the sense of equilibrium (statocysts). In the majority of cases, however, the receptor cells are either not organized into sense organs (they are evenly distributed over the body surface) or the organs are characterized by a very simple structure. The very structure of receptor cells is also highly primitive, e.g., the sensory cells, when stimulated, carry impulses to nerve cells through their own processes. The same applies to effectors: these are primarily epitheliomuscular cells that join the functions of epithelium and muscles as the very name suggests.

In free-living flatworms (planarians), a loose network of nerve cells covering the entire body is locally concentrated into primitive cords and ganglia. Roundworms are characterized by a much more developed concentration of nerve cells. In annelids, the nervous system takes the form of a ladder, where the head ganglia, linked by the throat ring, are connected to two cords that give rise to segmental ganglia linked with a commissure. In arthropods (crustaceans, insects, arachnids)—the descendants of annelids—the concentration of ganglia along the ladder is growing. The ganglia merge into larger and larger clusters of nerve cells (as the body segments merge). Insects are equipped with a complex organ of vision: compound eyes. In cephalopods (the group of mollusks that reached the highest level of development), the concentration of ganglia in the head is so great that they can be said to have a "brain." It is covered with a type of cartilaginous "skull" and its internal structure is complex. One can identify there particular centers, e.g., groups of cells that are responsible for closing the suction cups. Cephalopods are also equipped with the eye, the structure of which resembles closely the eye of vertebrates. In chordates (the group includes vertebrates), one can trace a wide range of brain evolution stages, from the extension of the front end of the neural tube to the human brain. One can also observe the evolution of receptors here, from cells randomly distributed over the body to such complex systems as eyesight in birds and primates or hearing in bats (that functions as an element of the echolocation system).

A detailed discussion of the evolution of the anatomical structure of the nervous system (and of receptors and effectors) would not contribute much to the understanding of the growing complexity of the worldview developed within the neural network. Therefore, I will not continue the above discussion. Instead, I will indicate the general trends in this evolutionary process. First, the system undergoes quantitative growth, i.e., growth in the number of nerve cells in the nervous system (accompanied by a growing specialization and functional efficiency of the cells). Second, particular neurons concentrated into trunks and ganglia, and finally into the brain. Third, the brain differentiated into a variety of areas responsible for separate tasks, which was accompanied by a relative growth of the size of the areas responsible for cognitive and decision-making functions (this is clearly visible in the evolution of mammals, primates, and man).

As far as the evolution of receptors is concerned, one can see a gradual development in the number, diversity, and complexity of sense organs (in purely anatomical terms, evolution proceeded from single cells to complex structures). The system of effectors, i.e., primarily the muscle system, has undergone a process of differentiation, from the stage of evenly distributed fibers in coelenterates and a uniform muscular sack in worms. The sack—that constituted a single big muscle oriented along the body axis (occasionally with circumferential fibers)—"broke up" during evolution into groups of muscles and single muscles that supported particular body segments, limbs, the head, the mouth apparatus, etc., while muscles got oriented in different directions adequate to the performed tasks (the evolution of the muscle system proceeded along with the evolution of the skeleton). The more complex the locomotor system (muscles, elements of the skeleton) was getting, the more subtle the movements it was capable of carrying out, which is exemplified best by the human hand. As mentioned before, the development of the system of receptors and the nervous system would make no sense without proper development of the effector system. Thus, all three are crucial in conditioning the development of the worldview possessed by a given organism. It should be remembered that the integrative and associative structures (described in this chapter) that constitute this worldview are always aimed at transforming the data from the environment into the behavior of an animal, irrespective of the complexity and indeterminacy of the transformation and the level of its mediation. Yet, in my opinion, it is the representation of the external world within the neural network of the human brain, created for purely biological (instrumental) reasons, that constitutes the content of our consciousness and it totally determines and limits the cognitive abilities of man. I will discuss this in the next chapters, where the evolution of the worldview will be traced in greater detail from the point of view of subjective mental phenomena—the "obverse" of the above-discussed neurophysiological processes.

5.

INSTINCTS, EMOTIONS, FREE WILL

W hat is the purpose of the brain—or the nervous system—from the biological point of view? There is, of course, one answer, as in the case of all organs and functions of living organisms, namely, the purpose consists in maximizing the chances of survival and reproduction. The only "purpose" or "sense" is that imposed on biological organisms by natural selection. This fundamental mechanism of evolution dictates the evaluation of various forms of adaptation, and it provides the undisputable point of reference in the evaluation of the survival efficiency of particular living organisms in their environment. The function of the brain of a given organism consists in generating the behavioral patterns of the organism that would meet the expectations of its selfish genes, i.e., the pursuit of reproduction in a possibly large number of copies.[1]

How can genes force brain to function in this manner? After all, the great majority of neural connection patterns in the brains of humans and higher animals is not innate, but it is acquired during ontogenesis, through the processes of learning and gathering experience. What are the factors that force the neural network to develop in the biologically "correct" direction during ontogenesis, and not in a random one? The answer is: the innate instincts (expressed in the form of certain neurophysiological mechanisms) and the reward-punishment system built into the brain. These mechanisms guarantee that the associative structures formed in the brain will produce an image of the world that would be as adequate with respect to reality as possible (meaning: instrumentally most efficient), and a system of behavioral reactions that aims at survival, acquiring food, and reproductive success.

The principal goal of each set of genes—duplication in a possibly high number of copies—is excessively general and abstract as a directive governing the development of the neural network within the brain during ontogenesis. Therefore, the goal got broken into a series of "subgoals" (during the biological evolution of the brain), namely: (i) avoidance of extermination (e.g., by predators); (ii) acquiring food; (iii) finding a sexual partner and reproduction; (iv) getting to know both particular features of the environment, where an individual lives, and the general features and rules that govern the external world (this is useful for the realization of the remaining subgoals); (v) ensuring the offspring gets proper care, food, and training; (vi) avoidance of illnesses and injuries that lead to the deterioration of an animal's condition; and so forth. Particular (broadly understood) primal *instincts* are responsible for the realization of these subgoals by the brain (and thus, for controlling the process of developing appropriate patterns of synaptic connections during ontogenesis), namely: self-preservation instinct, hunger instinct, cognitive instinct, parental instinct, instinctive avoidance of illnesses and injuries, and so forth. This list of instincts is, of course, incomplete and the demarcation lines between instincts are not sharp, so that the instincts have been differentiated in a slightly arbitrary manner. The cognitive instinct has a special place among these mechanisms, as its primary goal seems to consist in facilitating the realization of the remaining instincts (in animals; in humans it has been sublimated to a great extent to form a "pure" curiosity of the world, scientific interests, philosophy, etc.). This instinct is important for further considerations and I will keep returning to it.

Instincts are innate neurophysiological mechanisms based on appropriate neural structures that provide a system of reference for both the behavior of an animal and for the development of associative structures in the brain that are responsible for behavioral patterns. The hunger instinct is satisfied when the receptor of glucose concentration in blood registers a high level of the sugar, while the relevant mechanical receptors in stomach walls register pressure caused by the presence of food. The satisfaction signifies that the neural circuits related to the instinct are appropriately activated, and afterward ... but I will discuss it a bit later. The self-preservation instinct is not satisfied in the direct presence of a predator, which results in alarming all the functional brain centers that are correlated with this instinct. The satisfaction of the sexual instinct is, of course, related to having sexual intercourse, while the satisfaction of the "instinct" of avoiding illnesses and injuries consists in the avoidance of situations in which an animal is prone to develop an infection, to become ill from poisonous food (e.g., poisonous plants), or to injury. The cognitive instinct is related to the spontaneous brain activity aimed at the creation of new memory records, i.e., at the development of the system of associative structures. In order to satisfy it, it is necessary to collect a daily portion of new experiences. The portion is relatively bigger at an early age, when the learning processes develop in the fastest manner (this is one of the reasons for the propensity to play at young age). As mentioned before, the satisfaction (or its lack) of each of the instincts is reflected at the neurophysiological level in the activation (or its lack) of an (innate) functional structure of connections within the neural network that provides the basis for a given instinct. The brain is informed about the satisfaction or lack of satisfaction of a given instinct by the sensory data (properly processed) generated by receptors.

Particular "instinctual neural structures" send information about their activation or its lack to the central "evaluating module" in the brain, namely, to the *reward-punishment system*.[2] As the name suggests, the system operates according to the carrot and stick approach. There arises a fundamental question, namely: which is the donkey, whose behavior is managed by means of a stick and a carrot? The answer is simple (to my mind): the establishment of such-and-not-other associative structures corresponds to the behavior of the proverbial donkey. The term "associative structures" gets enriched here with one more meaning, for the reward-punishment system strengthens (makes more "patent")—by increasing the weight of appropriate synaptic connections (decreasing the levels of their excitation thresholds)—these associative structures, the establishment or activation of which coincides in time (or circumstances) with the satisfaction of an instinct (instincts), while it weakens (blocks)—by reducing the weight of appropriate synaptic connections—these associative structures, the establishment or activation of which coincides with the lack of satisfaction of an instinct. There is no mystery in it, to my mind. The reward-punishment system, the ramifications (axons) of which reach all the parts of the brain cortex, simply sends (through appropriate neurons/synapses) activating signals to the associative structure that is to be fixed, or inhibiting signals (through other neu-rons/synapses) to the structure that is to be weakened or totally eliminated. (It is also possible that the excitatory/inhibitory sig-nals are sent to the entire brain cortex, but only the recently acti-vated associative structure reacts to them.) It is also the reward-punishment system that decides which temporary "tentative" associative structures in the short-term operating memory will be transferred to and fixed in the long-term memory.

The reward-punishment system participates in the creation, removal, and modification of associative structures, primarily in the decision-making part of the brain (cortex), but also in the sensory and locomotor parts. Let us consider several examples. A small child realizing the cognitive instinct attempts to grab different objects or to walk. The combinations (random at first) of muscle contractions that result in a success are "rewarded" by the consolidation of the corresponding associative structures. When an insectivorous bird gets poisoned as a result of eating a characteristically colored insect species, the "instinct" to avoid illness and injury will dramatically weaken the associative struc-ture in the bird's brain that "associates" the species with some-thing edible and launches a specific behavioral pattern. A young animal is at first urged to run away from any moving object. Only by observing its mother does the animal "teach" the rele-vant associative structures in the brain which objects should be feared. As the neural network develops, e.g., in a growing mammal, the processes of "rewarding" and "punishing" apply to more and more complex associative structures that are respon-sible for proper recognition and understanding of more and more complex objects and laws of the real world, as well as for more and more complex behavioral patterns. For instance, the neural network of a predator will consolidate such hunting strategies (including right prey identification, sneaking method, distance assessment, the choice between leaping and chasing) that are associated with the successes in former hunting attempts. In general terms, the learning process is carried out by trial and error (additionally, the mechanism of imitating other individuals of the same species may be used). This is similar to the "learning" process in artificial neural networks, where the so-called backward propagation of errors corrects particular weights of connections by intensifying or weakening them, depending on the degree that the system reaction deviates from the expected reaction, so as to obtain an "output" reaction to an "input" stimulus that is as close to the expected reaction as possible. The process is of course much more complex in the brain.

It is difficult to univocally identify the entire reward-punishment system within the brain. It is commonly associated (at least its reward-related component) with the so-called dopaminergic system, i.e., a network of neurons with axons—ramified all over the brain—that release the neurotransmitter called dopamine. It is known that dopamine is released in the circumstances where instincts are satisfied (satisfaction of hunger, realization of sexual intercourse, removal of threats or stress-generating stimulus), which is accompanied in humans with positive experiences (to be discussed in a while). The narcotics that generate increased amounts of dopamine, e.g., amphetamine, produce ecstatic experiences, while dopamine deficiency leads to depression. It is not quite clear, whether the dopaminergic system is responsible for all the functions of what is customarily called the reward-punishment system. The brain of a rat, for instance, contains an area that is occasionally referred to as the "pleasure center." If electrodes are placed in the area and the switch is connected to a pedal in a rat's cage, the rat will stop being interested in food, sex, or any other activity and it will keep pressing the pedal until it starves to death. By introducing the electrodes into the "pleasure center," we create a powerful "artificial instinct," the satisfaction of which stimulated the reward-punishment system in a much more powerful manner than the satisfaction of any other instinct, which results in a tremendous intensification of the associative structure that generates the directive: "press the pedal irrespective of any other circumstances." A certain equivalent of the "pleasure center" is found in man (by stimulating different parts of the brain with electrodes and asking the patient about his experiences) in the so-called septum (a part of the brain cortex), while the "pain center" is sometimes identified with the so-called amygdala, although the evidence is less univocal in this case. I believe that it is safer to say, at the present stage of the development of our knowledge, that although certain aspects of the reward-punishment system have been discovered, many detailed questions remain to be elucidated.

The satisfaction or lack of satisfaction of instincts stimulates (positively or negatively) more than just the reward-punishment system. Such circumstances as sensing danger or chasing after prey activates the noradrenergic system (related to the transmitter called noradrenalin) that is also referred to as the energy system of the brain. While the dopaminergic system activates selectively only the "advantageous" associative structures, the noradrenergic system activates nonspecifically the majority of the brain by adding (in numerous neurons) an additional excitatory signal (see chapter 2) to the signal summed at the axon root, which generally speeds up the operation of the brain as well as the speed of its reactions to a particular situation. The stimulation of the noradrenergic system results in increased excitation, attention, excitement, and tension, while its deactivation leads to decreased tension, drowsiness, and slackness. It is clear that, in general, the noradrenergic system is activated in the case when one or more instincts are not satisfied, and it is switched off in the case when they get satisfied (the mechanism is a bit more complex, though, e.g., if an instinct chronically lacks satisfaction, it can lead to depression—decidedly a "low-energy" state). The serotonin system, responsible for controlling mood, is the third system, the operation of which is superimposed over the operation of the other two systems. The noradrenergic and serotonin systems also participate in the regulation of the sleep cycle.

Let us return, however, to the reward-punishment system, because we have reached a very important point in our considerations. The mental correlate of positive/negative stimulation of the reward-punishment system resulting from the satisfaction/lack of satisfaction of instincts and the corresponding intensification/weakening of the relevant associative structures (e.g., by activating the dopaminergic system and releasing dopamine) is found in the subjective experience of pleasure/pain. The reader could guess this when I mentioned the pleasure and pain centers a few paragraphs before. On the other hand, I believe that one can talk literally about the pleasure/pain center only in the creatures that are endowed with self-consciousness, and thus, with subjective mental states. It is difficult to decide, whether humans are the only creatures on Earth that are endowed with consciousness, but it is certain that our consciousness is developed in an incommensurable manner compared to animals.

One of the basic principles of psychology states that people are motivated to carry out actions resulting in positive (pleasurable) sensations, while they avoid actions associated with negative (painful) sensations. Thus, the reward-punishment system is a motivation mechanism inherited from animal ancestors devoid of consciousness, while its operation was adapted by the subjective mental sphere as pleasure/pain. Originally, the pleasure-pain axis reflected and corresponded perfectly well to the axis of adaptation-misadaptation (to environmental conditions) determined by natural selection. Do these axes still converge and point in the same direction?

In my opinion, they decidedly do not. I believe that the axes diverged considerably in highly developed human societies. Due to the prosperity achieved in such societies, and the corresponding absence of the direct need to struggle for survival, natural selection has become considerably weakened or even eliminated (except for trivial cases of lethal mutations), while the empty space of behavioral patterns has been filled with cultural selection and the whims of the "pleasure center." The latter, when freed from the direct supervision of biological evolution, tends first of all to satisfy its own needs, oftentimes in an extremely mediated manner (including both orgiastic debauchery and religious asceticism, as well as such "sublimation" of the cognitive instinct that turns it into a goal in itself, as is the case in pure science). What is more, after the emergence of the mental and cultural level of reality, the pleasure-pain axis oftentimes opposed the adaptation-misadaptation axis (i.e., their corresponding goals were incompatible), and thus, it overcame—at least partially—the rigorous dictate of genes over our behavior. This conflict between nature, on the one hand, and culture and the psyche, on the other, is due to the fact that the cultural level of reality—just like the mental one—generated its own, arbitrary objectives and laws, as well as mechanisms required for their realization that are absent from the biological level.

I will give a number of examples to translate these abstract considerations into concrete facts. A hermit does not hand over his genes, neither does he help his relatives to do so. The same applies to a priest, scientist, or artist, who decides not to have offspring, and do so in the name of realizing "higher" goals. By saving a person who is not our kin (or a representative of another race)—which seems obvious from the point of view of ethics—we support the genes of a potential competitor in the struggle for survival (competition to control limited resources of the environment) and often put our lives in danger. Other examples: having sex for nonprocreative reasons (contraceptives hold one of the leading positions on the "black list" of natural selection), alcohol, drugs, tobacco, unlimited pleasures of the palate. They bring about nothing but harm, from the biological point of view. First of all, such activities involve absurd allocation of energy, but also lead to venereal diseases, AIDS, obesity, hepatic cirrhosis, addiction, heart attacks and tumors (to name just a few). All the above examples share one characteristic feature: direct or more indirect stimulation of the "pleasure center" (i.e., the reward system) in the brain (if the stimulation is more indirect, as in the case of an ascetic, scientist, or artist, we can speak of "higher," more sublime human motivation). However, nobody would be able to convince me that this type of motivation pays at the level of natural selection. Anyway, behaviors incompatible with natural selection are found not just among humans: even elephants and apes get drunk with accidentally encountered fermented juice, because it stimulates their pleasure centers (from the point of view of evolution, the stimulation of the reward system in this case is "inconsistent with it principle objective"). A random and sporadic event of the type that eludes the mechanisms of natural selection becomes an immanent component of human societies (I am talking here about activities that are incompatible with the interests of biological evolution in general, and not only about drinking alcohol), where natural selection has become weakened by the "buffer of the civilization." It should be noted that the mechanism increases the genetic load of the members of society, and this state cannot be expected to last forever, since it does not seem to be stable.

The above conclusions apply directly to sociobiology—the scientific and philosophical view that human behavior (of both individuals and entire societies) is a derivative of biologically (and thus, evolutionarily) conditioned patterns that have been inscribed into our genes. Do I agree with the view? It depends on how we understand the rather vague concept of being a "derivative." If it means that human behavioral patterns are evolutionarily derived from behavioral biological factors that have been present in our lives until now, then the answer is yes. However, if sociobiology is understood to claim that all human behaviors with all their aspects can be reduced exclusively to biological conditioning factors, then I oppose such view. The case of sociobiology is similar to that of reductionism: both views have a profound sense unless they assume extreme forms. Doubtless, a living organism (including humans) is (also) a certain complex configuration of atoms, and no *vis vitalis* or spirit has anything to do here, while the life-related features gradually emerge from the processes occurring in inanimate matter during the ascent to the successive levels of the hierarchy of complexity, as it is claimed by a reasonable version of reductionism. However, if we follow extreme reductionism and claim that man is nothing but a certain configuration of atoms, and nothing interesting can be added here, then we reach the wall of the absurd. The language of physics, or even biology, lacks the terms and concepts to describe completely human beings and human societies. Both extreme reductionism and extremist sociobiology simply do not take into account the objective (in my opinion) status of the phenomenon of complexity (we are getting back to the main idea of this book), the real existence of certain levels of reality located above the physical level in the hierarchy, namely, of the biological level as well as mental and cultural one (I do not intend to treat these levels as separate entities). Extreme sociobiology is a nonverifiable myth,[3] and as such it has no place in the domain of science. On the other hand, I totally approve of sociobiology understood as a research program aimed at showing the extent of the impact of biological conditioning factors on our purely "humanistic" values and behavioral patterns, including e.g., ethics. In general, I consider myself to be a follower of moderate reductionism (identical, apart from the terminology, with rational holism) and a sociobiologist, yet I claim that these views become their own caricature if followed with sufficiently stubborn consequence.

I talked above only about feeling pleasure and pain. Yet the whole gamut of related feelings is much broader and it covers a variety of emotions, such as joy, the sense of triumph, satisfaction, fear, aggression, anger, sadness, depression, etc. What is the relation between pleasure/pain and emotions? Emotions[4] are (in my opinion) simply pleasure and pain with different situational connotations entangled in a variety of contexts. Positive emotions (joy, satisfaction) are of course accompanied by the sense of pleasure, while negative emotions (fear, sadness)—by the sense of distress. As it has already been mentioned, pleasure and pain constitute a mental correlate of appropriate stimulation of the reward-punishment system (primarily the dopaminergic system) in the brain. However, the system of psychic energy is also important for emotions, as it has impact on the general excitation of the brain that is controlled by the above-mentioned noradrenergic system. Some emotions—both positive and negative, for instance joy, the sense of triumph, fear, aggression (rage)—represent states of high excitation of the brain, while others—including satisfaction, sadness, depression—are correlated with a low level of "psychic energy." The two axes—pleasure-displeasure and excitation-lack of excitation—constitute the main factors that differentiate emotions (their subjective "coloring" and quality), while other, additional factors (in general: the situational context) are responsible for more subtle differences between emotions.

I have indicated above that the reward-punishment system and the innate instincts that operate through the system have significant impact (through selective consolidation/weakening) on the creation and modification of associative structures in the neural network. As these structures are components of the view of the external world in our brain, it becomes obvious that instincts are responsible for orienting the view and making it *intentional.* It means that our representation of reality is not neutral or impartial as, for instance, a photograph, an operational program of a robot or a perceptual pattern created in artificial neural networks that have been taught to recognize various visual patterns. On the contrary, the reward-punishment system establishes a definite axis that organizes all the activities undertaken by the organism, controlled by the nervous system. Therefore, I claim that the reward-punishment system constitutes the crucial element that makes it possible for an intentional "ego" to emerge in an animal organism equipped with the reward-punishment system. This is equivalent to the emergence of the *subject*—the *reference point* for all objects (things) of the external world. In the majority of animals, the "ego" of course lacks awareness of itself—it does not receive signals about its own existence. I believe that it gained full self-awareness—i.e., the ego started to perceive the fact of its own existence—only in humans. However, for the "ego," the intentional subject to be able to reach self-consciousness, it must first come into existence as a complex aimed (at least originally) at the evolutionary success, consisting of the reward-punishment system controlled by the instincts (that provide it with orientation).

Thus, the processes of decision-making and planning are

controlled by associative structures, the establishment and modification of which is supervised by the reward-punishment system that, in turn, expresses the will of instincts conditioned, in the final instance, by genes. Decisions in their essence are nothing but an anticipation of a behavioral strategy that will lead to possibly intense positive stimulation of the reward-punishment system (i.e., to possibly strong excitation of the "pleasure center" and possibly weak excitation of the "displeasure center"). It should be remembered that in the case of humans the reward (i.e., the neurophysiological correlate of pleasure) can consist not only in simple stimuli, such as the pleasures of the palate and sex, but also in carrying out scientific research or philosophical reflection (that satisfy pure cognitive curiosity), aesthetically oriented activities, ethical feelings, religious and mystical experiences, etc. This is one of the foundations of the nature of humanity. And this is all: every human act aims at more or less mediated (veiled), positive excitation of the reward-punishment system.

There arises immediately the problem of (conscious) free will—a phenomenon that has been dear to philosophers, ethicists, theologians, and lawyers for ages, and that is accepted by simple people in daily life. Unfortunately, both the logical analysis of the concept of free will, the current general knowledge of neurophysiology, and experiments carried out on volunteers seem to indicate unequivocally that there is nothing like free will (in the philosophical sense of the term). More, it does not even make sense—this is an empty and self-contradictory concept. I will try to present arguments in support of this emphatic thesis that may probably be outrageous for many people.

The problem of free will—its existence was considered to be obvious, and it was ascribed to the immortal soul as its immanent attribute—emerged naturally in the confrontation with pre-twentieth-century physics (headed by Newtonian mechanics), totally deterministic in its essence. The so-called Laplace's demon, an imaginary creature that has all the information on the position and momentum of all atoms in the universe, would be capable of recreating with infinite precision the entire past of the universe, as well as to predict its future. As the human brain is constructed of the atoms of matter, its operation, including the decisions it makes, should also be completely determined and planned in advance from the very beginning of history. Therefore, (at least some of) the advocates of free will were enthusiastic about indeterministic quantum mechanics in which chance is an immanent component of the behavior of atoms. Similar warm reactions were generated by the chaos theory. It claims that even in the case of strictly deterministic systems, their evolution is totally unpredictable in the long run, unless one has an infinite amount of information at one's disposal.[5] In my opinion, the optimism was due to a total misunderstanding of the problem. How could chance or chaos get us closer to free will? In its essence, it is a total antithesis not only of determinism but also of random or chaotic behavior. Free will refers to a spontaneous (occurring "by itself," whatever it means) and purposeful planning of actions—not reducible to the laws of physics—by an autonomous subject. Do dice or chaotic Brownian motions contain more free will in themselves than a tram moving on its rails?

It is often claimed that our consciousness is endowed with free will that guides our behavior. Consciousness, however, is not a separate entity. It is an epiphenomenon, a "side effect" of the operation of the brain and it is not capable of influencing the brain, the atoms in it, the transmission of impulses through neurons, etc. (it does not have any means, in the form of a physical factor, mechanism, or force, to achieve this) that it would have to do, since the functioning of the neural network is exclusively responsible for our actions. Consciousness may be treated figuratively, as a kind of passive screen that shows a projection of subjective correlates of the current activities of a certain fragment of the neural network in the brain. Consciousness emerges when we move from the functioning of single neurons to the functioning of organized networks made of huge amounts of neurons, just like the arrow of time in thermodynamics (and the resultant irreversibility of phenomena and the tendency of entropy to grow), absent in a system consisting of one or two atoms and emerging gradually, when we get to ten, hundreds, thousands, millions, billions, etc. of atoms. Both consciousness and the arrow of time represent emergent properties, the existence of which is objective to the extent that we are willing to ascribe to the phenomenon of complexity.

The relation of consciousness to the brain can also be compared to the roar that is a "side effect" of water movement in a waterfall. However, if the analogy is treated too literally, one can suggest that the sound wave may bounce back from a rock and disturb the water movement in the waterfall, and thus gain an influence on it. Therefore, it seems more accurate to treat consciousness as something so subtle, elusive, and nonmaterial as the beauty of a flower. The beauty emerges during the blossoming, but it should be clear that the beauty cannot in any way influence or distort the development of the flower. (The sense of beauty is equally subjective as the sense of having free will.) Thus, the relation between the brain and consciousness seems to be definitely unidirectional: the brain "generates" consciousness, but consciousness guided by "free will" is *a priori* unable to influence the brain, or our behavior through the brain. This leads to the conclusion that conscious free will is only a subjective, introspective sensation.

By its nature, free will is defined as a conscious phenomenon. This generates another problem, as physiology has shown convincingly, that the decision-making processes in our brain are unconscious in their majority (this is rather obvious: as I have already said, the processing of information in the brain proceeds in a parallel manner, while consciousness is treated as an entity, so that its successive states proceed one after another, in a serial manner). An illustrative example can be found in the so-called blindsight: a patient with a damaged visual route in the brain claims that he does not perceive visual stimuli (they do not reach his consciousness), while various reactions of the patient show undeniably that certain parts of his brain can "see" and guide his behavior adequately! This is due to the fact that the signals from visual receptors do reach the decision-making and locomotor parts of the damaged brain through "old" (in evolutionary terms) routes, inaccessible for the neural network that constitutes the "carrier" of consciousness. Another example is found in the so-called hemispherical neglect (resulting from certain cerebral hemorrhages), when a patient is not conscious of and cannot focus his attention on the left side of reality, although some of his subconscious reactions indicate something else.

Anyhow, not only the decision-making process itself ("free will"), but also the majority of our intelligence, thinking, and planning is carried out through unconscious processes. Today, this view is commonly accepted by neurobiologists. I know from introspection numerous cases, when a long-sought solution to a scientific problem appeared in my consciousness at a moment when I was thinking about something else (most often, when I was daydreaming).

Neurophysiological experiments provide more concrete and undeniable arguments against free will. The Libet experiment—which will not be discussed in detail, as its logic is quite complex—shows that the process of becoming aware of perceptual data lasts for about 0.5 second, while numerous "conscious" reactions to stimuli (i.e., reactions perceived subjectively as controlled by our conscious free will) appear after a much shorter time. We are dealing here with "secondary rationalization" of reactions that occur at the unconscious level, without any participation of free will, so that they appear—*post factum*, in introspection—as sovereign, independent and autonomous decisions of our "ego" (whatever it means).

The Kornhuber experiment is even more impressive. In this experiment, a patient was asked to flex his finger voluntarily at a consciously selected moment, after a period of motionlessness. During the experiment, the patient was monitored to record the electroencephalogram of the process. Of course, anyone can check it by experimenting with himself: a split of a second (virtually a twinkle of an eyelid—I know, because I checked it) passes between a conscious, voluntary decision to flex a finger and the actual flexing of the finger. The problem consists in the fact that increased activity of appropriate brain areas was observed not less than a second and a half before actual flexing of a finger! "Something" in the patient's brain took the decision to flex the finger over a second before the decision reached his consciousness. What's more, the patient's consciousness assumed the decision to be his own—an expression of his sovereign free will! The researcher carrying out the experiment could observe the encephalogram and his consciousness "learned" earlier about the decision-making process than the patient's consciousness. Thus, the researcher's consciousness knew in advance, what the patient's "free will" was going to decide in a while! Pardon me for the number of exclamation marks, but the experiment is truly fascinating and thrilling.

The above-presented considerations lead to the conclusion

that conscious free will is neither conscious nor free. Consciousness does not participate in the decision-making process, but it learns *post factum* about the decisions taken at the unconscious level. At that level, the processes of thinking, planning, and decision making consists in the activation, development, and modification of a complex of associative structures in the neural network of the brain, while this process consists in gathering experience and is channeled by the reward-punishment system, controlled by instincts. Not only is conscious free will unnecessary to explain the above phenomena, but there is simply no place for free will in these phenomena.

This will end my description of the neurophysiological operation of the brain, i.e., the functioning of single neurons and of a neural network consisting of neurons and connected to the system of receptors and effectors. In the following chapters, I am going to present the "other side of the coin," i.e., the psychic level (of subjective mental phenomena). I will return to neural networks for a moment in chapter 7, when describing the neurophysiological correlate of self-consciousness.

6.

THE NATURE OF MENTAL OBJECTS

I n chapter 4, I described the relational (connotative) functional structure of the neural network in the human brain.
Since I assume that the human psyche (consciousness), together
with all the mental objects it contains, *emerges* in some way from
the functioning of the brain, the structure must condition the
nature of the mental sphere in general, and of the contained
objects in particular. In this chapter, I will present my views on
the nature of particular mental objects, namely, concepts (and
the conceptual network), thought processes and sensations (and
some others: recollections, dreams, etc.). I will also discuss the
relation between the conceptual network and language. Some of
these issues have already been touched upon in the previous
parts of this book.

In my opinion, the structure and functions of neural connections in the brain find their mental correlates (reflection,
expression) in the *conceptual network*, consisting of particular
concepts. As I understand it, a *concept* refers to anything that can

(even potentially) become the (temporary) content of consciousness: that is, everything that the mind can somehow perceive, think, imagine, feel, explain, generate in the form of hallucinations, etc. Concepts can be very concrete, e.g., a tree, a stone, or very abstract, e.g., justice. They may refer to concrete objects, e.g., Julius Caesar or this and not other specimen of a red rose, to categories of objects (a planet), to unreal beings (an angel) and emotions (the feeling of fear). Finally, concepts may refer to "objects" so vague and undetermined that we do not find any languistic names for them. Concepts are not only "singular entities," such as words in a language, but also more complex entities that correspond to entire sentences and ideas. Concepts comprise even the most vague senses that appear in our consciousness. According to the above-formulated definition, when speaking about consciousness or the psyche we speak about systems of concepts.

Primary and secondary concepts (I will explain this categorization later) constitute the mental correlate of the formerly discussed associative structures of the lower and higher order. Only concepts (together with sensations received from receptors corresponding to activated integrative structures and processed by the "interpreter" of the conceptual network) can potentially become the content of the subjective mental zone, i.e., of self-consciousness.

I am aware of the fact that my broad definition of the term "concept" may not agree with its widely accepted usage. The narrower meaning of the term corresponds approximately to elements of the semantic memory, but I would like to extend it to cover other types of memory available in our consciousness, and especially the episodic memory. This is justified by the fact that the semantic memory is superior, in an important sense, to the episodic memory, and it conditions the general understanding of the records stored in the latter. I believe this approach to be highly justified, as all mental phenomena, such as sensations, thoughts, recollections, emotions, or dreams constitute nothing more than different forms of activation of various areas of the conceptual network[1] (I will discuss this promptly). The network is an interpreter of signals originating from the external world and provides them with a reference framework, i.e., it is the network that makes it possible for the signals to be understood. The inverse relation also obtains: the received sensations co-condition further development of the conceptual network. What is more, concepts are superior to language names (see below). The conceptual network constitutes the structure, the substance of sensations, thoughts, and other mental objects. As such, it deserves a privileged position in our considerations.

What are the consequences for the nature of concepts and the conceptual network from the general functional structure of the neural network in the brain, discussed in the previous chapters? The first consists in the fact that particular concepts within the network, just like particular nerve cells within the neural network, signify only by *connotation*, i.e., concepts signify in relation to other concepts. Thus, the sense or the meaning of a given concept is determined only by its reference to other concepts. There is no discrete, one-to-one, and absolute assignment of a concept to its designation (an object it refers to). To the contrary, the conceptual network (and the network of associative structures at its foundation, as mentioned before) "entwines" the facts (aspects) of this world during its development, just as a spider web may entwine the surface of a sculpture. In the last instance, each concept within the conceptual network is defined by all the other concepts, i.e., the entire conceptual network constitutes the proper context for each concept. In short: what a concept means results from its relation to other concepts.

As a consequence of the above, the conceptual network is—generally speaking—a continuous entity, i.e., concepts smoothly merge into each other (or they can do so, at least potentially). This is also due to the structure of the neural network, if only one assumes (which seems obvious) that a single concept is a derivative of connections between multiple nerve cells, which allows one to disregard single neurons (that operate in an analogue and not a discrete mode anyway) as located below the "resolution" of mental phenomena.[2] Thus, the conceptual network is totally different from language that obviously consists of discrete, "quantized" names. This does not prevent concepts from having the attribute of individuality. It is analogous to hills in a landscape: although there are individual hills, there are no strict borders between them. Two features are decisive for the identity of a concept. The first is the degree of concept determination, the degree to which the sense of a concept is singled out, the intensity of its meaning in the "semantic field." This feature determines the clarity of a given concept in our consciousness and the degree to which the concept seems self-evident and univocal to us. This depends on the *number* of relations between a concept and other concepts within the conceptual network. The second feature is the *quality* of the relations of a concept, i.e., which concepts it is connected to, which concepts it semantically refers to (and how), what other concepts define a given concept. As mentioned before, this feature is decisive for the meaning of a given concept. It should be noted that the first of the features of a concept is, in a sense, secondary with respect to the second feature. If it is known, what other concepts are referred to by a given concept, the number of relations of this type are also known.

As the conceptual network is a derivative of the neural network, it also contains—just like the neural network—both the innate component and the component acquired during ontogenesis. The innate component consists of the general predispositions of the brain to create concepts, and of the centers of their crystallization (the primordial meaning axes, to be discussed later on). On the other hand, particular concepts are developed during ontogenesis, since their "substance" is—at least partly—sensory in nature.

One can venture the following metaphor: various concepts are located in various places within the "semantic space" with a potentially infinite number of dimensions. The dimensions can be illustrated as "meaning axes" that establish a kind of Cartesian coordinate system within the space. In such space, concepts are developed by the "polarization" along the axes with respect to other concepts. In other words, the "semantic void" is laminated into particular new concepts. Here are examples of simple meaning axes: "big—small," "light—dark," "pleasant—unpleasant." Like concepts, the meaning axes would be (potentially) continuous entities, both along the axes and between the axes, when differentiating one from another. More, the type of polarization (the opposite axis ends) are also determined by concepts. Here, the connotative character of the conceptual network is manifested in its full nature. Concepts are determined by meaning axes, while the axes are determined by concepts. The introduction of the idea of axes is aimed primarily at elucidating the very idea of the conceptual network. The idea of axes, however, is important, as we humans have certain innate meaning axes (that I will discuss together with the evolution of the conceptual network during ontogenesis) that are used as the first interpreter of incoming external stimuli, and of the concepts based on these stimuli. These axes correspond, at the neurophysiological level, to instincts and innate integrative structures (of lower order). I'd like to remind the reader that the idea of meaning axes and the semantic space is nothing more than a useful metaphor that is aimed at intuitively elucidating the essence of what the content of our psyche is, since I do not intend to create a new philosophy and to bring new entities into existence.

As I have said, the conceptual network reflects the (functional) pattern of connections between nerve cells within the neural network and especially the system of associative structures inscribed into the network. At a given moment, however, only a small part of these connections is "realized," which means that only a small part of the processes of nerve cells and synapses conduct impulses. Moreover, the majority of processes in our brain are of unconscious character. If it is so, however, only a small part of the conceptual network may become the content of consciousness at a given moment, which is perfectly understandable when the huge size of the network is taken into account. At a given moment, the light of consciousness covers only those concepts that are currently "activated," and they are done so in a particular manner that allows them to become the content of consciousness (see chapter 7). Particular concepts that are covered by consciousness at a given moment can also appear in the consciousness with different intensity—this is due to the phenomenon of attention.

Different fragments of the conceptual network (and the neural network at its basis) may be activated in different ways. If they are activated by receptors (i.e., by signals from the external world), we are dealing with *perceptions*. Perceptions (based on the activation of integrative structures at various stages of integration) are understood and interpreted, and they have a sense only because of the manner in which they activate certain concepts of the conceptual network (and so, the corresponding associative structures in the neural network). The perception of the color red[3] is generated by the activation of the concept of red color, and the perception of a cat by the activation of the corresponding fragment of (the associative structure) of the neural (conceptual) network. There is nothing like a simple, elementary, atomic property of redness[4]—it is a concept that has meaning only by connotation, by reference to other concepts. Concepts used as interpreters for incoming perceptions are acquired during ontogenesis on the basis of (*inter alia*) the sensory stimuli gathered in the past, and of certain innate meaning axes and integrative mechanisms. Incoming perceptions are referred to existing concepts and potentially may be introduced (after proper processing) into the conceptual network as new concepts. The conceptual network totally determines the content and meaning of incoming perceptions. Perceptions are strictly related to the activation of integrative structure, the stimulation of which activates, in turn, the most concrete concepts (associative structures of lower order). Therefore, perceptions are characterized by "suggestiveness" and "clarity."

Having the above in mind, there is no sense in speaking (as many philosophers do) about "elementary perceptions" or *qualia*, such as the perception of blueness, for instance. The perception of blue color is "blue," because the neural structure (a group of neurons functionally connected in an appropriate manner) corresponding (equivalent) to the concept of blue color is connected to (it is a bit more complicated in reality) the light-sensitive cells in the retina that are sensitive to the blue section of visible radiation. During our life, the structure "learned" (by multiple excitation caused by various combinations of signals— see the mechanism for establishing associative structures) that signals from "blue" receptors are received in the case of a group of objects (i.e., there is a coincidence of blue color with certain objects, for instance with the sky over our heads). Then, our conceptual network assumed that "blueness" is a shared property of objects that stimulate "blue" receptors (and only "blue" ones). Therefore, "blueness" co-defines all blue objects in the neural (and conceptual) network, while blue objects co-define the blue color. One would not be able to understand or even "see" blue color without all these blue objects that one has encountered in his lifetime and introduced (as their representations) into his neural network. Therefore, blue color (its perception) is not in any sense a simple, independent, and autonomous entity. What we perceive (by means of our eyes) is electromagnetic radiation characterized by a certain wavelength. We do not perceive "blueness" in its immanence (as nothing like that exists), but only the core of the concept of blueness, when a corresponding fragment of the neural network is activated by signals generated by "blue" receptors (or, less intensely and clearly, by impulses generated internally by our brain, when it is dreaming, thinking, imagining, or recollecting what I will discuss promptly).

The very essence of the perception of blueness is determined exclusively by the neural (and thus conceptual) context of the concept of blueness. We do not experience blueness, because "blue" receptors (and then, neurons that mediate in the transfer and integration of information received from receptors) send different neural signals than "red" receptors (in fact, these signals are identical), but because the neural structure corresponding to blueness during ontogenesis has been stimulated by blue objects that activate "blue" receptors. The concept of blueness simply remains in different relations with the rest of the conceptual network (and the corresponding integrative and associative structures are in different relations with the rest of the neural network), compared to, for instance, the concept of redness, and this is precisely the essence of blueness. Let us imagine that the neural connection between "blue" receptors in the retina and the center of "blueness" are cut in a patient's brain, and the "blue" center is connected to "red" receptors (light-sensitive cells—cones—sensitive to red color). Then, we show the poor patient a red rose. What can he see? A *blue* rose, of course. This is so because the subjective sensation of blueness does not correspond to what stimulates "blue" receptors (the electromagnetic waves with right length), but to what activates the center of "blueness"[5] in the brain. What is "blueness" then? This is nothing more than a property shared by all objects that stimulate "blue" receptors during brain development, reflected in the brain in the form of the "center of blueness" (and the corresponding concept in the conceptual network). The "blue-nesses" of different persons (their subjective perceptions of blue color) differ slightly, but this cannot of course be verified (there are much more significant differences between normal people and color-blind persons). The "bluenesses" of a human and a fly are most probably almost totally incomparable (a fly has totally different receptors and a much smaller neural network with a different pattern of connections between neurons). Why do we experience blueness and a video camera does not (although it "records" blueness in some way, as we do)? This is so, because a camera does not know the concept of blueness, defined by millions of other concepts that could be activated by "blue" stimuli, and therefore it is unable to interpret and understand these stimuli as blue. I will return to the nature of perceptions further on in this chapter.

The question of *qualia* is tackled in the following thought experiment (that is supposed to demonstrate their independent and autonomous existence, irreducible to their material "base"). A woman neurophysiologist named Mary is an outstanding researcher dealing with colors, their broadly understood nature, their perception, their physical base—in short, with everything that science can tell about colors. Since her birth, however, Mary has been kept in an isolated room, where everything, including her body, was painted white, black, or grey in different shades. In other words, since her birth she had access to "black and white TV" instead of "color TV," and the TV was extended to cover all the world. During her research, Mary acquired complete knowledge of, for instance, red color, i.e., she knew the length of electromagnetic waves corresponding to the color, and she knew which cells in the retina are sensitive to the color and how nerve impulses are generated (and in which neurons) in answer to the perception of redness, etc. Mary, however, does not have the slightest clue how it is when one receives a subjective psychical sensation of redness. She does not know what sensation is generated when the "red" *qualum* appears in the light of consciousness. Therefore, when we allow Mary to leave her black-and-white room and enter the colorful world, the first perception of redness will provide her with some totally new knowledge that she has not been able to acquire by analyzing the theoretical and physical foundations of the perception of red color. This is supposed to prove that the understanding of the mechanisms generating some sensory perceptions is not identical with the experienced directly, subjective "content" of such perceptions.

The above-described thought experiment is an interesting case because its conclusions are correct (if they were not, subjective sensations would turn out to be *identical* with the objectively existing matter), while its premises and the argumentation are logically defective. In fact, it shows that philosophers failed blatantly to understand the operation of neurophysiological mechanisms. Can Mary "see" redness, when she leaves her black-and-white room and enters the colorful world? Of course, she cannot! She will not see redness, because she failed to develop the *concept* of redness during her life: she lacks the associative structure corresponding to this concept. This leads to the lack of ability to differentiate and "understand" redness. What will Mary see? Of course, she will see the right shade of grey, corresponding to the degree of saturation of red color, just like in the black-and-white television. This grey shade, however, will be—at least initially—indistinguishable from the grey shade corresponding to green color with the same saturation. Is Mary going to acquire the ability to differentiate and perceive colors and to develop the concept, for instance, of redness? It depends on the flexibility of her neurological mechanisms at the age of maturity, on the condition of her light-sensitive cells, responsible for detecting colors (whether they have degenerated or not), etc. It seems justified to conjecture that she will never acquire the ability to "see" color efficiently. This is suggested by multiple neurophysiological observations and experiments. If a cat is raised since its birth in a room with only vertical strips (everything painted accordingly), its ability to perceive horizontal strips, lines, and contours will be highly impaired when it is released into the normal world. A grown-up person who has been blind since birth and whose ability to see is restored, sees only an incomprehensible chaos of sensations (inspiring the feeling of incomprehensible terror) instead of normal images. During their development, children can learn their first ethnic language (learning other languages is much easier because they find their place in the already-shaped "language structures" within the neural network) only during a certain period (up to the sixth year of age, if I remember correctly). If they do not, they will not be able to learn any language, which is shown, for instance, by the commonly known case of Kaspar Hauser.[6] Thus, it is most probable that Mary will not be able to acquire full efficiency in perceiving colors. The fact that the argument based on Mary's paradox is totally pointless (although its general conclusions happen to be right) is one of numerous examples showing that ignorance and the lack of understanding of scientific achievements lead philosophers astray.

Thoughts appear as a result of activation of concepts in the conceptual network, which is due to the internal, autonomous activity of the brain. In this case, primarily the "highest" areas of the brain cortex (the prefrontal cortex in particular) are activated, i.e., those areas where the operating memory is located. Thoughts can be (and often are) modified and disturbed by perceptions (that's why we try to cut off signals from the external world when we want to concentrate). On the other hand, the complete cutting afferent stimuli off may lead to serious impairment of psychical functions (there are relevant experiments that show this), including the process of thinking. This shows that perceptions have a very important role in supporting and directing the stream of consciousness even if one tries to concentrate at a given moment on thinking—an activity that is so autonomous and seemingly so cut-off from the external world. The concepts activated during the process of thinking are in principle very general concepts (located at the top of the integrative-associative hierarchy) that lack the directness of perceptions. Therefore, thoughts seem to be much more "distant," elusive, and vague compared to perceptions.

The nature of other mental objects (within consciousness) —recollections, dreams, hallucinations, emotions, etc.—can be characterized in a similar manner (within the framework of my theory). They all constitute cases of activation of various concepts in the conceptual network, and the differences between them result from the activation of different areas of the conceptual network (and the corresponding neural network) in different manners (e.g., intensity is a relevant parameter). So, *memories* are connected primarily with the activity of the brain cortex areas that are responsible for collecting memory records (e.g., temporal lobes), mainly in the episodic memory. Dreams involve the operation of various centers in the brain, but their content is not "channeled" through the stimuli from receptors (as in the case of the stream of consciousness in the waking state), yet only a small part of dreams can become conscious (remembered) after awakening anyway. In the case of hallucinations, the crucial role is played by an "unnatural" (i.e., not caused by signals from sensory organs) activation of the sensory cortex (for instance, due to the use of narcotics that distort the functioning of neurotransmitters). Emotions, in turn, constitute the expression of the stimulation (or its lack) of the motivation mechanism (punishment/reward) in the brain (the dopaminergic system constitutes its part), and the pleasure and pain "center" in particular. (While emotions are generated in subcortical regions, the processes occurring in the brain cortex are responsible—in my opinion—for their subjective content). It should be remembered, however, that (according to the concept presented here) all these types of mental objects—when they appear in the stream of consciousness—constitute certain sets of concepts that signify by connotation, and not some autonomous entities signifying by themselves, or some qualitatively incommensurable monads.

This is the right place for an explanation of the mutual relation between the conceptual network and language. Unlike the conceptual network, which is in principle continuous, language consists of clearly isolated, discrete names. According to my theory, discrete language names correspond to the clearest and most univocal concepts in the conceptual network[7] (i.e., to the concepts characterized by the highest intensity of the "semantic field," due to numerous relations with other concepts). Vague and hardly graspable concepts that are difficult to interpret do not have their language correlates. All emerging ideas elude language, before they are formulated/expressed through a rich conceptual framework. Thus, the extent of the conceptual network is larger than that of language. While a concept can be ascribed to each language name, no names can be ascribed to poorly determined concepts that are nothing more than mere hints of senses. Moreover, language itself—which seems so autonomous and well defined (isolated)—also consists in its deeper layer of concepts that support language names. This should be understood in the following manner: both the symbol (e.g., a sequence of letters in a word), its meaning (a set of senses designated by the symbol), and the correspondence relation between them (i.e., the symbol and its meaning) are limited to (constituted within) the conceptual network. This becomes obvious automatically, when one remembers the neurophysiological foundation of language, namely, the relevant fragment of the neural network, where neurons—operating according to a fuzzy logic—signify only and exclusively by connotation. Thus, the conceptual network is a more general and primordial structure than language. The latter constitutes, in a certain important sense, only a part of the former. Therefore, concepts and not language names and sentences constitute the content of our consciousness. The entire process of understanding language proceeds exclusively at the level of the conceptual network. Language was originally used for communication between two consciousnesses, two conceptual networks. The emergence of language obviously facilitated the very process of thinking, since language is a perfect instrument to manage the entire conceptual network. This is probably the origin of the theories of "language-limited thinking" (e.g., Wittgenstein's philosophy). Anyway, it is probably impossible for higher forms of thought and consciousness to arise without language. It remains a fact, however, that concepts—superior with respect to language names—constitute the primary substance of psychical phenomena.

To sum up the above considerations: in my theory, all "momentary" mental objects that fill in the current stream of consciousness—such as sensations, thoughts, recollections, dreams, or emotions—are simply various forms of activation of the relevant fragments of the conceptual network (and the neural network at its foundation). Having clarified this issue, I would like to focus on the structure and properties of the conceptual network itself, and on its evolution during the ontogenesis (during individual development) and phylogenesis (within the biological evolution) of humans.

Let us commence with ontogenesis. At the moment of birth, the conceptual network of a human being has an incipient form. This is not unexpected, since an embryo receives only few external stimuli during uterine life, e.g., the mother's heartbeat. An infant, however, has certain innate functional neural connections (patterns of connections between nerve cells) that give form to the integrative structures, and thus to sensations. They predispose the infant, for instance, to organize received stimuli into spatial, temporal, and causal patterns, etc. Thus, certain basic methods of organizing sensations into the image of the external world are predefined in advance (not to mention about such fundamental, already discussed mechanisms of integration of visual stimuli, as the perception of movement, lines, contours, and colors). The newly emerging conceptual network (based on the sensory data) also contains other developmental nuclei—crystallization centers. These are certain primordial meaning axes defined by purely biological meanings and values (thus, they could be identified with instincts). The following examples of axes illustrate the case: "satiation-hunger," "heat-cold," "sense of safety—lack of safety," "cognitive inquisitiveness—its satisfaction." At any moment, the signals transmitted from receptors are received by the relevant (functional) centers in the brain, where the circumstances are interpreted as they arise, with respect to the above-mentioned axes. The centers promptly inform the (already discussed) pleasure/pain center in the brain about the outcome. The center, in turn, generates the adequate reaction of the infant. If it is satiated, feels safe, and its cognitive instinct is satisfied, the infant is calm. Otherwise (e.g., when a toy that it is interested in is taken away) it cries. However, the pleasure/pain center is used not only to guide the current behavior of the infant. It is also responsible for associating neutral external stimuli with innate instincts, which means it leads to emotional assessment of these concepts (into positive and negative ones). The child learns (by establishing appropriate associative structures) which behaviors lead to satisfying its instincts, and which stimuli (objects, properties) originating in the external world are correlated with positive states, and which are not. Thus, the innate integrative mechanisms and meaning axes differentiate the signals from the environment (or rather, from the sensory organs), organize them appropriately, and segregate into certain categories that constitute the nuclei of future concepts. Generally speaking, the integrative structures give form to newly emerging concepts (associative structures), while instincts lead to their evaluation. The differentiation and organization in question proceed along the already existing meaning axes. The newly emerging concepts, in turn, provide the foundation for new semantic axes.

As the conceptual network develops during ontogenesis, new secondary meaning axes are established on the basis of the already-existing axes. By induction, the successive, regularly repeated sets of stimuli are ordered with respect to the axes as new concepts. This is the way the representations (corresponding to associative structures of lower order) of concrete "real" objects of the external world are created in the conceptual network. I will use the term "primary concepts" to refer to such concepts that correspond directly to simple "facts" in reality. General, abstract concepts—secondary concepts—are created (at a higher level) in a similar way, as the primary concepts are, when the mind "perceives" numerous similar sets of detailed concepts of lower order (these can be both primary concepts and secondary concepts with a lower position in the "hierarchy of abstraction"). Autonomous thought processes play a significant role in the formation of secondary concepts. The processes consist in the active association of various associative structures, the identification of concurrence, analogy, and repeating patterns in these structures, in the extraction of various regularities and rules, and finally in their consolidation in the form of associative structures of higher order. General concepts, which correspond to such structures, do not refer directly to the "elementary" aspects of the external world, but rather to various relations between different sets of such objects. They contain less "sensory substance" (originating mainly in the processing of receptor signals by the relevant integrative structures) than the primary concepts. They have a higher admixture of the "subjective component"—a derivative not of the external world (seen through the prism of senses), but of the neurophysiological mechanisms of the functioning of the brain.

To sum up: a child is equipped with only the basic semantic axes that have purely biological (instinctive) significance, and the general mechanisms of ordering and integrating signals from the external world. These are genetically "implanted." The first concepts arise on the basis of sensory perceptions that passed through the interpreter consisting of these axes and mechanisms. The development of the conceptual network during ontogenesis is based on several basic principles. The following are the general regularities in the development of the conceptual network: existing concepts "laminate" into more detailed concepts; new concepts are introduced into the conceptual network by induction, repeated perception of certain sets of sensations or detailed concepts; finally, the determination of existing concepts increases (they become clearer). These processes are strictly connected. In principle, they are different manifestations of one and the same process.

One could venture to arbitrarily differentiate the five main aspects of the evolution of the conceptual network during ontogenesis: namely, quantitative growth, the appearance of a certain surplus with respect to the purely "straightforward" representation of the world, learning, mastering language skills, and the emergence of self-consciousness.

The fact of quantitative development of the conceptual network since birth to maturity does not seem to require further demonstration. As I have just mentioned, an infant is equipped only with a few innate semantic axes and the general mechanisms for integrating sensory stimuli, while the abundance of the conceptual network of an adult is known to everyone from introspection. As far as the surplus with respect to the "straightforward" view of the world is concerned, I am referring here mainly to the appearance of secondary concepts as a result of autonomous information processing of "higher order" that occur in the brain—i.e., to thinking.

Since practically the entire conceptual network is acquired during ontogenesis, it must be created in the process of learning. There are many ways of learning: by observation, by imitation, by the method of trial and error, and finally by means of linguistic messages, both oral and written. The availability of learning methods is conditioned by the degree of development of a person's conceptual network. An infant, with his conceptual network in a nuclear form, is able only to observe passively the surrounding reality. Elements of the reality gradually acquire different meanings depending on their relation to the realization of purely biological instincts (satiation of hunger, sense of security, etc.). Having acquired a certain orientation in the world, one can start to experiment with it, e.g., by moving itself or shifting various objects (the part of the conceptual network related to the locomotor system develops at this stage). This facilitates the development of a spatial view of the world and the identification of its properties. Learning by trial and error is in fact a method of active perception, where an individual observes his initially random actions and the corresponding reactions of the world (the cognitive instinct is the driving force here). No wander that an infant shows interest in each new toy and obligatorily puts every object within its reach into its mouth (taste and smell are very primeval senses, both in terms of biological evolution and of ontogenesis). A more advanced method of acquiring knowledge consists in imitating parents or other adults (mirror neurons participate in this process) that requires a certain degree of understanding of the world, i.e., a relatively well-developed conceptual network.

The efficiency of learning processes increases extremely once an individual masters the skill of using the system of artificial symbols—language. Language—while it (or its semantic layer) is a part of the conceptual network—facilitates immensely the efficiency of using the network. Language names facilitate the "identification" of the best defined concepts that have corresponding labels in the form of names. Syntactic structures are useful for proper ordering of names (and thus, of the corresponding concepts) and for the creation of definite structures out of the names. Language, as a common social convention, allows a conceptual network of an individual to be translated into a conceptual network of other individuals. New pieces of information and the correlated concepts are quickly and efficiently positioned within the existing semantic structures so that there is no need to learn them by direct experience. This increases tremendously the process of learning, i.e., the process of developing a conceptual network. Language provides an additional advantage because it allows information to be transmitted at large distances, both in time and in space.

Each ethnic language not only stimulates the development of the conceptual network and the view of the world created within the network, but it also shapes the view to a large extent by its grammatical structure. The world has only one structure, but the structures of various languages differ, and often they differ dramatically. There are languages that do not have verbs (in our sense of the term), which means that the significance of temporality in their views of the world is different than in our language. Languages that contain names of only two colors or three numbers obviously condition the manner of perceiving the world. In still other languages, the names of objects are composed of their attributes, e.g., both a tree and a hand contain a component that designates branching. A different language structure involves a different logic of the language and the conceptual network (and of the world viewed through them). Therefore, the final shape of the conceptual network depends to a large extent on the language (and culture), in which it is created.

In an infant, self-consciousness (the subjective mental sphere) is present in a rudimentary form, or it may even be absent. Thus, consciousness must develop (emerge as an epiphenomenon from a particular manner, in which the neural network operates) during ontogenesis. This is of course a gradual process, and it is not possible to identify the moment of emergence of consciousness in a nonarbitrary manner. Anyway, it would not make much sense, as in the case of attempts at identifying the moment, when a seed transforms into a tree. In chapter 7, I will explain my views on the essence and the emergence of self-consciousness.

Although the "final" conceptual network (and the neural network at its base) of an adult human is in every particular case a result of all life experiences (sets of received stimuli)—which I described above—its general structure and developmental predispositions are genetically encoded. The latter features must have been formed as a result of gradual development of their primeval counterparts during the biological evolution of our ancestors, whose constitution was simpler and whose brains were less developed, which means that their conceptual network must have been less extensive and less complex.

As in the case of the development of the conceptual network during ontogenesis, one can indicate five aspects of conceptual network development during biological evolution: quantitative development, the emergence of a surplus quality, acquisition and development of various modes of learning, acquisition of symbolic language, and the emergence of self-consciousness. For an organism to have even the simplest forms of a conceptual network, the organism has to be equipped at least with the simplest form of a neural network. As mentioned above, such a network appears for the first time in cnidanans. Thus, such multicellular organisms as sponges or *Mesozoa*, as well as the entire world of plants, fungi, and unicellular organisms (protozoa, algae, bacteria) do not fall into the scope of the considerations on the origins of the psyche.

The beginnings of the conceptual network were extremely simple during biological evolution. The first conceptual networks were innate, and their function consisted in simple "recoding" of environmental stimuli into behavior. Thus, the conceptual network had a purely biological significance. The set of stimuli that our early ancestors reacted to and the repertoire of behavioral reactions were very modest. Their conceptual networks were characterized by a low degree of complexity, while the concepts themselves were simple and poorly determined (by the number of semantic interrelations with other concepts). Jakob von Uexkül indicated the tick as an example (it was not one of our ancestors, to be sure). Its entire life consists in climbing a tree and waiting (often for over a dozen years) for a passing animal, falling on the potential host, and finding a place on its skin where it can draw blood. For a tick, the concept of a "host" is limited to the smell of the butyric acid (the signal for the tick to fall down from its tree), the body temperature, and the taste of blood. The "world" of a tick, i.e., its conceptual network, is very modest. For a tick, the concept of a "roe-deer" is a combination of several simple sensory perceptions. For humans, the same concept (with the same designate in the animal world) is defined by a huge number of other concepts concerning the appearance and behavior of a roe-deer, its taxonomic position, its anatomic structure, its physiology and biochemistry (what may even include the fact that the behavior of atoms of a roe-deer is governed by quantum mechanics), and cultural aspects of the animal (e.g., hunting), etc. In the case of humans, practically the entire conceptual network consisting of concepts-nodes is the semantic context that "defines" the concept of a roe-deer. In the case of a tick, the network consists of just a few nodes. No wander that the "psyche" of a tick seems to be totally incomparable to the human mind. A similar degree of complexity of the conceptual network was probably characteristic for our early ancestors. Thus, there is no other explanation but that the human conceptual network (together with the extremely rich worldview) was established during biological evolution and developed gradually from some primitive conceptual network, comparable to that of a tick.

While the first and simplest conceptual networks were used primarily for automatic "translation" of environmental stimuli into animal behavior, the more advanced and developed conceptual networks acquired a certain "surplus" with respect to this function. The reaction to a stimulus became mediated and delayed because time was needed to "confront" a given stimulus with the worldview already established within the conceptual network, and to take the right decision. Thus, the stimulus-response reaction became (at least seemingly) less determined and the associative processes that correlated the data transmitted from receptors to the relevant sections of the conceptual network became the germs of thought processes. The latter became even more autonomous later on, as no distinct environmental stimulus was necessary to launch them. Autonomous brain operation aimed at planning future activities and at decision making was sufficient.

"Remnants" of various stages of conceptual network development can be found even today in humans and higher mammals. For instance, the unconditioned reflex in humans (e.g., automatic withdrawal of a limb after touching a hot object) was acquired very early during evolution. It can be found in the simplest organisms equipped with a nervous system (e.g., the sea anemone), while in humans it is "supported" exclusively by the spinal medulla, without getting the brain involved in any manner (of course, the brain and consciousness are informed about the incident *post factum*, when the reflex-governed activity has already been carried out). The conditioned reflex discovered by Pavlov constitutes a slightly higher form of stimulus-response relation. The scientist observed that if feeding a dog (that results in salivation) is always accompanied by a sound signal, then the signal itself will be able, after some time, to cause salivation in the dog. In this case, conditioning consists in associating a biologically neutral stimulus (sound) with a biologically significant stimulus (presence of food) and a purposeful reaction to this stimulus (salivation). This is a beautiful example of a simple associative structure. Humans can also be conditioned to carry out similar "involuntary" actions, but such phenomena can already be observed in very simple organisms, such as snails. On the other hand, the behavioral repertoire of the majority of animals does not exceed this level. The "surplus" of their conceptual networks is still of a rudimentary character. The "proper" thought processes (although it is difficult to trace a clear-cut border that would separate thinking from non-thinking), consisting in managing totally new situations and solving complex problems as a result of an autonomous brain operation that "models" various actions and anticipates their results, can be observed only in evolutionarily advanced animals (higher chordates and cephalopods, and in a more explicit form—in anthropoidal apes and dolphins). As mentioned before, thought processes require secondary concepts that reflect (within the worldview established in the conceptual network) the principles governing different objects of the external world.

The rise of the surplus functions with respect to the functions involved directly in transforming sets of received stimuli into behavioral patterns of an animal was inseparable from the change of the proportion between the innate behavioral repertoire and the behavioral patterns acquired during ontogenesis, from the integration and interpretation of perceptual sensations. The acquired patterns gained a dominant position in higher animals. This was due to the development of the processes of memory recording. The "final" shape of a conceptual network was gradually becoming less genetically determined and shaped to a greater and greater extent by learning processes, which considerably enhanced the dynamics and flexibility of the network. This was enhanced by the period during which parents care for the offspring, when parents function as a type of buffer, a behavioral protection of a young individual, which reduces environmental threats, provides the child with food and facilitates the process of acquiring experience (especially by direct transfer of information) until the young individual reaches maturity. It became possible for the offspring to be initially equipped only with the "innate" (genetically determined) nucleus of the conceptual network. The rest of the conceptual network was acquired during ontogenesis.

This method of conceptual network formation allowed the networks of higher animals to differ considerably from those of lower animals, equipped primarily with genetically encoded structures. First of all, a conceptual network acquired by learning could be much larger and much more differentiated, as the capacity of genetically transferred information—huge in itself—is smaller by many orders of magnitude than the information-carrying capacity of the brain of a mature mammal, not to mention the human brain. The human brain contains about 100 milliard neurons. Each of them can potentially connect (functionally, through a synapse) to every other neuron, which yields a virtually unimaginable, astronomic number of possible combinations of connections. The amount of information contained in the connections actually established in the brain of an adult human is also huge (I have carried out approximate calculations that yielded a virtually unbelievable number). If the DNA of all chromosomes of humans were used only to define interconnections between particular neurons in the brain (in reality, it has a wide range of numerous other functions to realize), only a very small part of the brain of an adult human could be encoded in this manner. This leads to the simple conclusion that larger conceptual networks can be established only by learning.

As mentioned before, the process of learning may be carried out by the trial-and-error method, by observation, by induction (identification of regularities in received sets of stimuli from the external world), or by imitation of parents and other mature individuals. The advantage of such learning (as opposed to the "learning" during biological evolution) consists in its speed (the entire process is comprised within the life of an individual), flexibility, and its surplus quality that allows individuals to react adequately to unpredictable situations that have never been experienced before, and to the characteristic features of a given environment inhabited by a given individual. Innate ordering schemes of neural (and conceptual) networks were also established by the trial-and-error method (during biological evolution), through mutations of the genetic record, what resulted in the rise of variants of conceptual networks, reflecting more or less adequately the biologically relevant aspects of the world. Then, these variants were screened by natural selection—only the best models could survive and undergo further evolution. However, this type of "learning" is very slow (acquisition of evolutionary memory is measured by geological time) and very ineffective, and the conceptual networks created in this way are inflexible and totally unreceptive to corrections during ontogenesis (thus, they cannot reflect short-term changes in the environment and the differentiation of habitats occupied by a given species). When it is not necessary for the entire neural (and conceptual) network to be genetically encoded, it becomes possible for the nervous system, especially the brain, to evolve much faster. Learning as acquisition of experience during ontogenesis enhances the shaping of thought processes as an autonomous activity of the neural network. The conceptual network established by learning is therefore much richer and more flexible; it evolves faster and is equipped with the possibility of autonomous operation, which allows it to represent the external world better than a conceptual network that is genetically preprogrammed.

In the recent history of human evolution, language—or, to be more exact, the cultural transmission carried out by means of language—played a very important role in the development of the conceptual network. Each ethnic language is of course only an element of a given culture, but it is certainly a distinguished element. First, as I mentioned before, language not only facilitates considerably the management of a conceptual network of a given individual, but also allows conceptual networks of different individuals to be translated (only approximately) into each other (i.e., it allows communication). Second, the grammatical structure of each language shapes to a certain extent the structure of the worldview in the conceptual network, and thus it also shapes the form of thought processes based on the network. Third, language is a new channel of information transfer parallel to the biological channel and able to operate vertically (from one generation to another) and horizontally (between individuals of the same generation). What language transfers is the very content of a given culture, or the "conceptual network of the culture," understood as a sort of Popper's third world. This allows a new community member to inherit the experience, inventions as well as myths and misconceptions gathered by generations of his/her ancestors. Thus, on the one hand, the knowledge of the real world and various practical skills are cumulated, while, on the other hand, "accidental" ailments of culture are developed, such as religion, rituals, art, etc.

The last aspect of the shared development of the neural and

conceptual networks during biological evolution consisted in the rise (in humans and, possibly, in some other animals, in a vestigial form) of self-consciousness, i.e., the subjective sphere of psychic sensations. Since it was a breakthrough that led to the rise of the third level of reality (following the physical and the biological ones), I will discuss it in a separate chapter.

7.

THE RISE AND ESSENCE OF SELF-CONSCIOUSNESS

First of all, I believe that consciousness—understood as a psychological phenomenon, as a subjective sense experienced by the subject that there exists some external object—is a derivative of self-consciousness, i.e., of the sense of one's own "ego" because it is the sense of one's own "ego" that is constitutive for the essence of a mental subject. I would also like to separate "mental" consciousness from "instrumental" consciousness. The two senses of consciousness are often mixed up. Psychological consciousness is a mental phenomenon that belongs to the third level of reality (apart from the physical and the biological ones) that I have already defined. As such, it is a special category. Instrumental self-consciousness, on the other hand, is simply something's awareness of something else, a representation of an object that is realized within a subject that realizes it. Psychical consciousness must, of course, comprise the instrumental consciousness, but—in my opinion—it must also comprise something else, an element that results in the emergence of the zone of subjective

131

sensations of our mind from the purely biological (neurophysio-logical) functioning of the neural network that forms its basis.

I would like to point out at this point that only the psychical approach to consciousness makes sense and is heuristically promising, while its instrumental understanding seems to be vague and leads to unnecessary terminological confusion without bringing about any serious results. I will explain why in a moment. What does it mean that a given "subject" that becomes aware of an "object" creates in itself an image or rep-resentation of the latter? The example of e.g., a red rose comes to our minds. We become aware of the rose by its appearance, smell, and touch. We find here an object existing in the external world and a reflection (more or less precise—it does not matter) of its presence in our mind, that is—in simple worlds—we are aware of the rose. We can take advantage of this fact by, for instance, picking the rose and giving it to some other being. Is it not enough to define consciousness satisfactorily?

Unfortunately, the issue is not that simple. The above defin-ition, based on the concept of representation, does not say any-thing about the absolutely most important question: why is the awareness of a red rose in our brain a *subjective psychic phenomenon* that is introspectively experienced as an element of our mind, and not of the external world (matter)? According to the former reasoning, the sensation of a rose is nothing but an appropriate stimulation of the concept of a rose in the conceptual network through stimuli from receptors, or—in the context of the neural network—the activation of an adequate associative structure. Why does a sphere of subjective sensations emerge, in this case and not in other cases, from the purely physical circulation of impulses in a network of nerve cells?

The instrumental definition of consciousness, based on the relation of representation, applies to many cases where we would never think of talking about a subjective psychic sphere. Let us take a camera as an example. According to the definition, a light-sensitive film is in a sense "aware" of a fragment of the external world that the camera objective is pointed at, because it contains a representation of the fragment in the form of an appropriate spatial pattern of grains of photographic emulsion exposed to light. Let us call this type of instrumental conscious-ness a static-passive consciousness. Everyone would agree that this is not the type of consciousness we are interested in, when considering the human mind.

Does the problem consist in the fact that the "conscious-ness" of a photographic camera does not take into account tem-poral changes of the image? Fine. Let's replace the photographic camera with a video camera that registers such changes. Let us call this type of instrumental consciousness a dynamic-passive consciousness. It is still far from our human consciousness.

A photographic camera and a video camera faithfully reg-ister certain aspects of reality, yet they do nothing with them, neither do they use them for any purpose. Perhaps conscious-ness emerges when its presence is a condition of producing a particular result and/or achieving a particular goal? Let us ana-lyze the operation of a fridge thermostat that regulates the tem-perature by using negative feedback. It registers the current temperature value (i.e., it is instrumentally aware of the temper-ature, while the temperature is represented here by an appro-priate curve of a strip made of two metals with different thermal expansion coefficients) and switches the cooling device when the actual temperature rises above the set temperature value. Let us say that this is a regulative consciousness. Again, it is cer-tainly not a satisfactory correlate of the human consciousness (as far as I know, a fridge does not experience subjective psy-chical states related to the actual temperature value).

Similar regulative consciousness, in a much more complex form, can be found in the case of "experiencing" the presence of lactose in the environment by a bacterium, followed by appropriate reaction to this presence (the above-discussed theory of operon). The appearance of this sugar triggers the entire chain of biochemical-genetic processes. The enzymes of the path of lactose catabolism are synthesized. Finally, the sugar is used as a building material and/or energy-providing substrate. The entire process is much more complex than in the case of a thermostat, but this is only a quantitative difference. As opposed to the maintenance of a constant temperature in a fridge, the use of lactose by a bacterium has a biological and evolutionary sense, as it is a purposeful process that increases the chances of survival and the speed of growth and multiplication of bacteria. Moreover, the entire network of regulative processes/mechanisms at the biochemical and genetic levels is aware of the presence of lactose. Although this seems to be a step in the direction toward human consciousness, based on a network (a neural one, and not a biochemical and genetic one) characterized by biological purposefulness and containing in its structure a certain (functional) reflection of certain aspects of the external world, we are still far away from psychic consciousness.

Perhaps it is sufficient for any neural network to become aware of something to give rise to subjective psychic sensations? It seems hardly possible for fundamental reasons. In a network of nerve cells, it is difficult to find something that would be absent from the biochemical and genetic network, while it would simultaneously lead to automatic emergence of the subjective sphere. If this is not considered sufficient, let us remember the example of cnidarians equipped with a very simple "nervous system" consisting of dispersed quasi-neural cells. This system is responsible for the polyp contraction as a result of touching an object. However, does the polyp really experience subjectively the presence of the object? Hardly anybody would agree with this diagnosis.

It seems equally improbable that the emergence of the subjective psychic sphere is related only to a sufficiently big size and complexity of the neural network. The problem consists in the fact that we are considering here only quantitative and not qualitative differences. The neural network of a frog—much better developed than that of an anemone—is certainly aware instrumentally of, for instance, the image of a fly on the retina of the frog's eye by activating the concept of a "fly" in the conceptual network of the frog. This situation seems quite analogous to the awareness of a rose gained by the human brain (at the beginning of this reasoning). Although we introspectively experience a subjective sensation of a rose, we are not inclined to grant this type of sensation to a frog. This is not due to the quantity, as it seems, for if we multiplied the number of neurons in the brain of a frog to reach the number of neurons in the human brain, the frog would most probably (at least in my opinion) not become "spiritualized."

The instrumental, representation-based concept of consciousness turns out to be unable to explain what seems to be crucial for human consciousness, namely, the existence of the subjective sphere. According to this concept, the consciousness of a photographic camera, a video camera, a bacterium, a sea anemone, a frog and a human being belong to the same category, while they differ by the degree of complexity, i.e., the differences between them are primarily quantitative. The uniqueness of human consciousness, however, demands to be explained. It seems intellectually unjustified to grant the status of consciousness in the remaining cases, as it introduces unnecessary semantic confusion. Therefore, the following conclusion seems unavoidable: Not so much a *degree*, but a *type* of complexity and organization of a conscious subject (its network of nerve cells) is crucial for the psychic level of reality to emerge from the neurophysiological functioning of the brain. There must emerge an appropriate SYSTEM. This assumption constitutes the basis of my concept of the essence of psychic consciousness and self-consciousness that I am about to present.

In my opinion, the relation of *self-orientation* (self-directing) or, more generally, of *self-application* constitutes this new quality, the specific mode of functional organization of the neural network that has led to the emergence of self-consciousness, and thus of the psychic consciousness. At the level of the conceptual network it was equivalent to the formation of a subject (within the network) capable of observing the processes occurring within the network, including the subject itself. In other words, it was equivalent to the focusing of the cognitive apparatus—hitherto focused on receiving the worldview shaped in the conceptual network—on itself, its own image in the network. This is the event that has led to the emergence of the third psychic level of reality, comprising the entire sphere of subjective sensations in our brain.

The passage should be understood in the following way: During evolution, the brain developed a certain cognitive and decision-making "center" (functional rather than clearly definable in anatomic terms) that compared data received from receptors with the memory records, coordinated various functions of the central nervous system and made decisions concerning the stimulation of effectors. The center (most probably dispersed across a considerable part of the brain cortex, and the prefrontal cortex in particular, as it is the seat of the "operating memory") confronts the signals from the environment with the already-existing conceptual network and the worldview shaped

within the network. At the level of the neural network, its operation consists in the activation, creation, and modification of
associative structures of the lower and the higher order. A fundamental role in its functioning is played by the autonomous
activity of the brain cortex, i.e., thought processes. If it is arbitrarily separated as a relatively well-isolated system, its inputs
will comprise receptors and the parts of the neural network
(primarily the sensory cortex that contains integrative structures) that process the data provided by receptors, as well as the
existing memory records. The emergence of psychic consciousness (equivalent to self-consciousness) would consist in orienting a part of the inputs of the system on the system itself, i.e.,
in self-recognition of the processes occurring in the above-mentioned center. The processes would also "process" themselves, as
they formerly processed the data obtained from receptors (or
from episodic memory records during sleep). *Apart from the image
of the external world, the center established within itself an image of
itself as well.* In other words, the center projects (maps) itself on
(in) itself and creates a model of itself within itself. This is
essentially a relation of self-application, analogous to a great
extent to the relation found in the liar's paradox,[1] the Russell's
antinomy, the Gödel's proof or in the logic that the "Catch 22"
is based on. The psychic correlate of this state of affairs consists
in the emergence of the concept of "ego" within the conceptual
network, what results in an apparent delamination of the entire
network into the "subjective" sphere focused on itself and on the
image of the external world (localized primarily in the prefrontal cortex), and the "objective" sphere that establishes the
image of the external world (located in the sensory cortex and
in the part of the temporal cortex that contains memory records,
especially episodic, but also semantic ones). *This self-orientation
(directing-on-itself) corresponds to subjective psychic sensations.* No

wonder that—in my theory—the psychic consciousness is closely linked to self-consciousness, as the former is conditioned by the latter. The *psychic* awareness of the presence of an object from the external world results simply from the act of projecting the "cognitive center" in the brain on itself (establishing a model of itself within the center), while the center is aware of the object in the *instrumental* manner. This explains why humans have psychic awareness, while a sea anemone and (most probably) a frog have not. The latter simply lack the relation of self-application. The difference between a brain (and a mind) without consciousness and a mind (brain) with self-consciousness is presented in Figure 11.

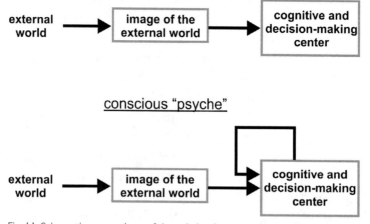

non-conscious "psyche"

external world → image of the external world → cognitive and decision-making center

conscious "psyche"

external world → image of the external world → cognitive and decision-making center

Fig. 11. Schematic comparison of the relation between the cognitive and decision-making center and the worldview in a non-conscious and in the self-conscious brain (mind). In a self-conscious brain, the cognitive and decision-making center receives its own image and the view of the external world through the same "channel" that the non-conscious brain uses only to receive the view of the external world. A self-conscious cognitive and decision-making center projects itself into itself, creating a model of itself within itself.

The "seat" of self-consciousness is located, in my opinion, in the prefrontal cortex, i.e., the place of occurrence of the phenomena at the base of the operating memory.[2] The processes of thinking, planning, and decision making that occur in this area are based on temporal, short-term associative structures (constituting a part of the short-term memory content) that can, but do not have to, be transferred to the long-term memory. Thus, one could venture to claim that consciousness is maintained by the short-term memory (although it also involves an activation of long-term associative structures). This would be confirmed by the fact that people with a damaged hippocampus—devoid of the possibility of acquiring records in long-term memory—still seem to be "equipped with" self-consciousness.

Above, I differentiated three levels of reality: the physical, the biological, and the psychic level, while renouncing any claims of their absolutization. In my opinion, the manner the biological level emerges from the physical level is analogous to the manner the psychic level emerges from the biological (neurophysiological) level; the emergence is based on the relation of self-application. I indicated above how this should be understood in the case of the emergence of consciousness. Now, I will discuss in brief what it means in the case of the emergence of the phenomenon of life.

From the point of view of cybernetics, a live organism may formally be considered—while abstracting from the material form—to be a hierarchically organized network of regulating mechanisms, primarily negative feedbacks. (Examples of such regulating mechanisms can be found in the above-discussed inhibition of a metabolic path by the product of this path or in the regulation of glucose concentration in blood by insulin and glucagon). The entire network of negative feedbacks, constituting the cybernetic identify of a given live individual, is aimed at one (evolutionary) goal: the reproduction of the organism, i.e., the multiplication of its identity (in this case: the cybernetic identity). This identity, constituted by the negative feedbacks, is oriented at itself, at its own propagation (this is certainly a case of the relation of self-application). On the other hand, the reproduction of live organisms is a perfect example of a positive feedback (e.g., when there are no environmental limitations, the number of bacteria on a medium grows from one cell in an exponential manner: we have 1, 2, 4, 8, 16, 32, 64... cells in the subsequent generations). Thus, a biological individual may formally be defined as a network of negative feedbacks (regulating mechanisms that ensure survival) subordinated to the primary positive feedback, i.e., to the self-replication of the network (equivalent to the cybernetic identity of the individual).

The analogy between a network of regulating mechanisms and the neural (and conceptual) network is rather strong. First, both networks (as I have mentioned before) constitute a certain functionally purposeful representation of various aspects of the external world important for the survival of an organism (the neural network is essentially a subsystem—with a high position in the hierarchy—of the entire set of regulating mechanisms of an animal). Second, the significance of particular elements of a network—negative feedbacks, on the one hand, and neurons and associative structures (concepts), on the other hand—is established by connotation, i.e., an element makes sense only with reference to other elements and finally in the context of the entire network. I have already explained what it means in the case of concepts. In the case of regulating mechanisms, it is clear that they do not operate in a void, but rather some of them are aimed at the realization of others, and in general terms: all the mechanisms are aimed at the realization of all other mechanisms (this applies to all goal-oriented processes occurring in live organisms). For instance, the maintenance of a constantly high level of ATP (by the relevant negative feedback) makes the synthesis of proteins possible (the process uses ATP). On the other hand, some of the proteins function as enzymes in the metabolic path of ATP synthesis, and their concentration is also maintained at the appropriate level by the relevant (genetic) regulating mechanisms that—in turn—require proteins and ATP.... Of course, the entire network of interrelations between various regulating mechanisms in a live organism is much richer and more complex than it could be demonstrated in such a simple example.

Thus, one could venture the following general thesis: a "higher" level of reality (biological or psychic) emerges from the "lower" level (physical or biological, respectively) by self-orientation (the relation of self-application) of a network of elements (negative feedbacks or concepts) that signifies by connotation and is an evolutionarily purposeful representation of certain aspects of the external world. In this formulation, the emergence of psychic consciousness (self-consciousness) from a network of neural connections in the brain devoid of consciousness would constitute, in a sense, a "repetition" at a higher level of the act of emergence of the phenomenon of life from inanimate matter. Such formulation of the problem seems to agree with the spirit of Bertalanfy's theory of systems, although I must clearly point out that it is my own conception.

It follows clearly from the above discussion that I do not consider the three levels of reality that I have identified to be some separate beings in the philosophical sense. I think that a "higher" level emerges from a "lower" level as a result of a certain special type of relation between the elements of the lower level, and I identify this particular type of relation (in the discussion above). Therefore, I consider the dispute between moderate reductionism and moderate holism—concerning the question, whether there emerges "something qualitatively new" at the new level—to be void and pointless, as the answer depends exclusively on the adopted definition of what we consider to be qualitatively new. If we reject primitive reductionism (everything, including life and consciousness, is nothing else but a movement of atoms, and nothing interesting can be said on the subject) that ignores the problem of complexity, as well as primitive holism (life and consciousness are totally new beings, the existence of which is conditioned by vital force and spirit, respectively) that seems philosophically out of date, we can say that the new thing at the "higher level" is a special mode of functional organization of a system of elements (the relation of self-orientation) that are common components of the "lower level," but in "normal" situations they enter into other (in principle, less complex) configurations at this lower level. It is immaterial for the essence of the problem, whether these facts are called reductionism or holism. The SYSTEM is the key to the issue.

The above-proposed definition of self-consciousness does not, of course, allow for the existence of a philosophical, idealistically construed "free will." Yet, as I argued earlier, such free will is already excluded by a purely logical analysis, general neurophysiological knowledge, as well as by particular experiments (Kornhuber's experiment and Libet's experiment). Personally, I do not feel any psychological discomfort due to the absence of such idealistic free will. I leave such worries to philosophers and humanists. I believe that we still know too little about the physical reality, as well as about the functioning of the human brain, to cope with this problem (this is not hypocrisy playing safe, but rather an expression of epistemological modesty). For the moment, it is sufficient that the concept of free will functions well (and it should continue to do so) in the psychological, social, cultural, ethical, and legal contexts. Yet, at the current stage of development of scientific knowledge, the concept seems to be a philosophical hypostasis. This apparent contradiction should not be wondered at, as the content of the great majority of cultural senses cannot be derived from the laws of physics or biology that belong to the domain of natural sciences.

Finally, I would like to comment on the so-called quantum theories of consciousness (recently in vogue), forced by some physicists, e.g., Roger Penrose. They attempt to present consciousness as a result of macroscopic quantum effects, as a type of extremely "complex"—meaning no less than "variously enfolded" (!)—wave function. In my opinion, instead of solving the problem, this move evades it and pushes it to the lower (!), physical level of reality, where it can be hidden behind our ignorance (more complex wave functions are beyond our current calculation possibilities). For instance, quantum theories of consciousness do not answer the following question: what would be the difference between the hypothetical quantum consciousness and other macroscopic quantum phenomena, such as Bose-Einstein condensation. In my opinion, physicists show an extreme lack of understanding of the status (objective status, in my opinion) of the phenomenon of complexity (unfortunately, evolution shaped our brain in a manner that makes us perceive complexity as something not quite real), while the essence of consciousness should be looked for at a level higher than the biological (neurophysiological) one, i.e., at the psychic level of reality.

One more vital point should be underscored. My conception of the genesis of self-consciousness is not just another philosophical speculation. On the contrary, it meets all the conditions required to be met by a sound scientific hypothesis, namely, it allows one to formulate predictions that can be experimentally tested. Here are some of them. First, it follows from the conception that there are closed nueral loops in the brain of an adult human being that lead from the dispersed "cognitive center" in the prefrontal cortex and back to this center. Second, a temporary blocking of these circuits due to an external intervention leads to a temporary switching-off of self-consciousness. Third, such closed neural circuits are not found in the brains of infants (in the period covered by the so-called infantile amnesia). Fourth, they are not present in the brains of the majority of animals considered to be devoid of conscious-ness (beginnings of consciousness may probably be present in the brains of chimpanzees and dolphins). Fifth, the transmission of signals in the closed loops ceases during sleep, at least during the deep sleep phase. Sixth, lots of evidence indicates that there exists a minimum "atomic" moment of consciousness that lasts for about 0.5 second (compare, for instance, Liebet's experi-ment). It could coincide with the period of a single "cycle" of signals running in the postulated closed neural loops.

8.

ARTIFICIAL INTELLIGENCE

When discussing the question of the nature and the physical (neurophysiological) correlates of the human psyche and self-consciousness, one cannot fail to mention, at least briefly, the question of Artificial Intelligence. The problem concerns conditions to be met by man-made devices, e.g., computers, robots, or neural-like networks to become intelligent and to be endowed with a psyche. Both phenomena (intelligence and psyche) that are of interest for us would become considerably more elucidated if these conditions were identified.

The very claim that it is in principle[1] possible to create artificial devices endowed with the subjective sphere of psychic sensations has, of course, far-reaching consequences in the domain of outlook on life. For such a claim assumes tacitly the following: 1. no qualitatively different being (in the philosophical sense)—the spirit, for instance, is necessary for consciousness to emerge; 2. the physical correlate of consciousness is not found in a particular type of material (e.g., neurons constructed from proteins, lipids, etc.), but in a particular type of functional organization of a network of elements, constructed from practically any substance (metal, silicon, proteins), but remaining in appropriate formal interrelations (primarily informational and cybernetic). The reader might have already found out that I belong to the group of strong advocates for this option. However, there are also advocates for the claim that it is in principle impossible to create a *deus ex machina*, i.e., an artificial intelligence. Before presenting arguments in favor of this thesis, I will attempt to defined what one should understand (in my opinion) by the term "artificial intelligence."

The "natural intelligence," i.e., the human mental sphere, has multiple aspects or components. Among them, one finds intelligence in the commonly accepted sense, the psyche, self-consciousness, subjectivism, thinking, emotional sensations, and personality. One immediately faces the problem, whether all these aspects are necessarily correlated with each other (all of them are present in humans), or whether, to the contrary, they can be separated and exist independently, and if so, which of them can condition (to the greatest extent) the emergence of an artificial intelligence. After all, many modern computers and robots (as well as animals) behave in an intelligent manner, according to commonsensical canons. They may even "think," although one would not attribute self-consciousness, personality, and the subjective sphere of emotions to these entities. In my opinion, however, "true" intelligence and thinking consist primarily in posing problems and goals to oneself, and not only in solving external problems, imposed by others or evolution, as it occurs in the case of computers and animals, respectively. Therefore, I believe that all the above-mentioned aspects of the subjective mental sphere are mutually conditioned and cannot exist separately. For instance, autonomous thinking allows a goal to be realized, which is related to positive emotional sensations experienced by a subject endowed with self-consciousness and personality. The idea of creating a "pure mind" devoid of motivation (and therefore of emotions) and unaware of its own existence seems to be, in my opinion, a utopian chimera, impossible to be realized for fundamental reasons. This idea indicates a total misunderstanding of the essence of the base of the human psyche. Therefore, I consider the term "artificial intelligence"— suggesting indirectly the possibility of existence of such a mind—to be particularly infelicitous (but it has been accepted and therefore I have to use it). In my opinion, the emergence of self-consciousness is the necessary condition for the existence of true (psychical) intelligence. This diagnosis certainly does not come as a surprise for the reader who has become acquainted with the previous chapter.

Having determined approximately the meaning of the term "artificial intelligence," we have to identify the criteria that allow one to decide, whether a given device is endowed with psyche or not. The commonly known Turing test consists in exchanging questions and answers (e.g., written on sheets of paper) with an "interlocutor" who is not seen, because he is in another room. Our task consists in determining, whether the interlocutor is a human being or a machine. If we cannot univocally identify the "interlocutor" as a man or a computer, and it turns out it is a computer, this means the machine has passed the Turing test, and therefore it is endowed with a psyche and consciousness. The Turing test is the basis of Searle's thought experiment known under the name of "Chinese room" argument that is supposed to discredit the very possibility of constructing an artificial intelligence. Instead of a computer, there is a group of Chinese persons in the room. They do not speak English, but receive cards with questions in English. The

Chinese provide answers to these questions (on cards) following a set of instructions in Chinese (corresponding formally to the algorithm of a certain computer program) that define a set of purely mechanical rules concerning the transformation of sequences of English words in questions into sequences of English words in answers. If the set of rules were sufficiently rich, the group of Chinese would pass the Turing test (they would be considered to be a conscious system), although they would understand neither the asked questions, nor the provided answers. Therefore, a computer cannot be endowed with consciousness, as it transforms data in a purely mechanical manner, according to a predetermined set of rules.

I believe the apparent "paradox" of Searle's Chinese room is misleading and essentially unwise. First, one would have to show that the Turing test is actually a reliable criterion for identifying systems endowed with a psyche.[2] Second, it is not that obvious (at least for me) whether a room crowded with Chinese persons would pass the Turing test. Third, the replacement of microchips in a computer with Chinese persons in a room is as much a delusive trick as it is heuristically abortive. The fact that a conscious Chinese being does not understand English does not prove anything. For it is obvious that a single microchip of a computer (corresponding to a Chinese person) does not understand the operations carried out in a computer program. Nothing less than a highly complex *system* of microchips organized in a special manner can become the base of consciousness. The analogy to the human brain comes here naturally to one's mind. Only the entire system, and not any of its elements (neurons), can be conscious. Searle did not demonstrate that the system consisting of Chinese persons and the set of rules cannot *as a whole* be conscious and understand English (although it could seem strange). The psyche is an emergent phenomenon, and as such it requires the phenomenon of complexity to be granted objective existence. Numerous natural sciences, developed on the basis of this excessively primitive reductionism, still refuse to accept this requirement.

Another error in Searle's argumentation consists in attributing to artificial intelligence the principle of operation of a modern digital computer that realizes a strictly determined algorithm of a program and operates with a precisely determined set of discrete symbols and rules that govern their potential interrelations (i.e., their syntax). The Chinese room is after all an analogy to such computer. For this reason (and not for the reasons supposedly implied by the thought experiment in question) I believe that a set of translation rules followed by the Chinese persons, microprocessors or appropriately trained shrimps would not be able to become conscious. If it is so, then what is it that modern computers lack and future electronic information processing systems would have to acquire to come into possession of the psyche?

I will skip the obvious question of the size and the degree of complexity of the entire system. I will commence the next point with the statement that such a "computer" should operate as a neural network, in an analogue, approximate, and nonalgorithmic manner, contrary to its modern relatives. It should not be based on a discrete zero-one logic. Information should be processed in a parallel manner, while the processing system itself should be able to undergo modifications and to evolve (learn) just like modern neural-like networks. Particular system elements should not operate in a totally determined manner. They should learn their functions through the interaction with other system elements and from stimuli that enter the system from the outside. While modern computers "think" in the linguistic layer (this also applied to Wittgenstein, at least in his own opinion) by operating on discrete and defined symbols, ordered by a set of grammatical rules, a conscious computer would operate on the basis of a deeper conceptual layer (absent from modern computers), where particular elements would signify by connotations, as neurons do in a neural network. In a "conceptual network" of such computers language would be a secondary, high-order structure, nonautonomous as it would be semantically dependent on the network itself.

A computer endowed with a psyche would have to have a correlate of the motivation center, the "seat of drives," and of the cooperating reward-punishment system in the human brain.[3] The seat would consolidate these connections of neuron-like elements that would favor the realization of some autonomous goals, corresponding to drives (in live organisms, these are imposed by the logic of biological evolution), e.g., the cognitive drive or the self-preservation instinct, etc. All this would form a base for the subjective sphere of emotional sensations. It also seems necessary for such a computer to have a correlate of spontaneous, autonomous, cognitive, and associative activities constituting a correlate of thought processes.

Such a conscious computer would have an image of the external world, constituted by the connotation of its elements, constantly modified by new signals from the world (from which come sensations, concepts, ideas, and other subjective mental phenomena discussed above). The development and functioning of such world projection would be supervised by a certain cognitive center. It would be focused not only on the external world, but also on itself, which would lead to the emergence of a subject, the sense of its proper "ego" and of the entire subjective sphere of psychic phenomena. Only such a self-conscious subject would be able to formulate its own goals and problems, i.e., it would be endowed with true intelligence.

I keep having a preponderant impression that the entire system has to get entangled into contradictions for self-consciousness and personality to emerge. This refers not only to purely cognitive and logical inconsistencies, but also to conflicts between the cognitive drive and numerous other drives (self-preservation instinct, reproductive instinct), and between the adaptation-misadaptation and pleasure-pain axes. In humans, the accumulation of contradictions and conflicts has resulted from millions of years of brain evolution that added new components to the existing ones without any pre-established plan, in an opportunistic and sloppy manner from the point of view of perfect structural solutions (this has been discussed in chapter 3). However, there arises the following question: If optimum technical solutions had been implemented, would the human race consisting of individuals endowed with personality and subjectivity—whose minds are endowed with irrationality and continuous vacillation between different motivations as their immanent attributes—come into being? I fear that, to create a computer endowed with a psyche and personality, it would be necessary to "implant" the system of contradictions and internal conflicts (collected over centuries) in its structure.

To sum up this part of my considerations, I would like to indicate the three features that I consider crucial for the emergence of artificial intelligence (or rather, artificial psyche), namely: a relational network of information processing elements that signify by connotation, a motivating instinctual center (implying intentionality of the system), and a certain self-focused cognitive system. It is obvious that none of the currently existing electronic devices meets the above-indicated criteria.

9.

THE COGNITIVE LIMITATIONS OF HUMANS

If one should ask what we know with absolute certainty, an honest answer to this question would be: very little or nothing at all (at this high level of abstraction, it is impossible to refrain from philosophizing, whether one likes it or not). This diagnosis induces an impulsively strong resistance in many people. For, there is a huge edifice of scientific knowledge collected laboriously over centuries, and we have achieved incredible progress in the development of technology and medicine based on the primary sciences. It suffices to indicate our knowledge collected in everyday experience: one knows, for instance, that there is a person he knows in the same room where he is. Unfortunately, any philosopher could easily show that the thesis attributing absolute and certain character to this type of knowledge is invalid. It suffices to claim (as solipsism does) that the external world (together with other people "populating" it) does not really exist, but it is a figment of our consciousness or imagination, whichever one likes. Irrespective of a commonsensical critical evaluation of such an attitude and of the futility of philosophical disputes, one has to admit honestly that solipsism (just like numerous other philosophical systems) cannot be refuted in an undisputed and unshakeable manner.[1] Unfortunately, even scientific knowledge, no matter how impressive in its immensity and exactness (the latter is owed primarily to mathematics and to the methodology proper for sciences), does not have sufficiently unshakeable foundations to be considered utterly certain and absolute. The program of reductionism—which works perfectly well in explaining physical foundations of chemistry, biology, or even the psyche, in the last instance started to crumble when physics reached the lowest level of quantum phenomena that can be derived, to a large extent, only from the properties...of our mind. I will illustrate the issue by presenting shortly a history of views (maintained primarily by philosophers) on mutual interrelations between spirit and matter.

The relation of spirit (consciousness, mind) to matter (the external world, the objective reality) is probably the greatest mystery in the history of philosophy. The view known as materialism maintains that matter is primary with respect to consciousness, that it is a result or a "by-product" of the functioning of the human brain. This is, more or less, a view represented (or at least used) in the hitherto presented considerations in this book. Idealism, on the contrary, maintains that consciousness is the only truly existing being, while the so-called external world (or broadly understood matter) is only a product of consciousness, an area of the psyche isolated in a particular manner. It is true that not all variations of both above-mentioned philosophical systems pose the question in such a radical manner. Some of them only assume that one being is superior to the other. There are numerous alternative conceptions, each treating the problem in question in a characteristic manner. In spite of the diversity of proposed solutions, the mystery of the relation of spirit to matter has not been solved until now, while it has continued to instigate new difficulties. In this book, I present this question within the perspective of the biological sciences. For philosophy, however, this point of view is obviously one-sided to a great extent.

In the dispute in question, natural sciences will naturally tend toward materialism. The impressive successes of physics and biology in describing and understanding various aspects of the world allowed scientists to believe that scientific methodology constitutes an incredibly powerful tool of cognizance. The methodology is based, among others, on reducing the structure and functioning of complex phenomena and systems to a relatively small number of relatively simple principles that govern the behavior of the components of such systems. Oftentimes, these principles can be expressed quantitatively in the exact language of mathematics, while predictions made on the basis of such principles show superb conformity with observations and experimental results. The research program outlined above, known as reductionism, used to triumph in the history of the development of natural sciences.

Since the beginning physics excelled in reducing the entire wealth of the world to manifestations of simple interactions between elements of a few types. All "material" objects found in the universe, including galaxies, stars, planets, plants, animals, works of art and the brains of the masters who painted them turned out to be nothing more than conglomerates of atoms of several dozens of elements found in nature or of the three elementary particles: protons, neutrons, and electrons.[2] The image of the *Mona Lisa* reaching our eyes is nothing but a spatial distribution of quanta of electromagnetic radiation (photons) with different wavelengths. No reasons were found to claim that there is "anything else" that contributes to the structure or has impact on the functioning of the above-mentioned objects. What's more, the properties of macroscopic systems—for instance, molecules of chemical compounds, crystals, gases and stars, planets or even (in principle) living organisms—were managed in a great number of cases to be derived from the properties of atoms (that, in turn, result from the attributes of the three elementary particles). Detailed research at the biochemical, cellular, and physiological levels showed that no supernatural force, like *vis vitalis*, is necessary to explain the functioning of living organisms, with their complexity and purposefulness, and to breath life into a bunch of "inanimate atoms." It turned out that the need for such a breathing and absolute act is an illusion of the human mind, as life emerges slowly from the cooperation of atoms and organic macromolecultes made of atoms, as one gradually passes from the level of chemical compounds to that of metabolic pathways, cells, and organisms. Thus, there is no essence or core of the phenomenon of life that would assume the form of a categorically different being, as life is a derivative of a set of physical and chemical processes that do not differ in principle from the processes occurring in inanimate matter.

The concept of *vis vitalis* as an additional being that ensures exceptional character of life simply turned out to be a hypostasis. Does not one find it similar to the concept of the absolute spirit that is supposed to be the touchstone of the special status and total irreducibility of the psyche and consciousness to the biological level, and to the physical one, in the last instance? In fact, natural sciences, and neurophysiology in particular (which I attempted to demonstrate, incompletely as may be), made a huge step toward de-absolutizing the concept of the spirit as a separate being. First of all, there are no doubts whatsoever about the possibility of explaining the physical and bio-chemical mechanisms of the conduction of nerve stimuli as a polarization and depolarization of the membranes of nerve cell processes, and of explaining the phenomena occurring in synapses. Huge progress has been made in the research on the integration of sensory perception in the brain and in explaining the mechanisms responsible for the formation of memory records. Dramatic changes in the psyche, in characterological features, and various mental capabilities were observed to occur as a result of damage to particular brain centers. For instance, cutting the corpus callosum connecting cerebral hemispheres leads to the emergence of two independent and, to a large extent, autonomous consciousnesses in one head! Stimulation of various areas of the cerebral cortex allows one to generate a variety of visual, aural, and other sensations, or to bring remote memories to mind. We know that consciousness emerged grad-ually from the "psychic void" during biological evolution, and that it develops gradually during ontogenesis as well (neither our ancestors, amphibians, nor human embryos are endowed with it). Probable models of thinking were proposed, consisting *inter alia* in a rivalry between various activity patterns of the cerebral cortex for the available areas of the cortex (by analogy to the rivalry of living organisms for the available environmental resources). It will probably be possible to verify these models experimentally in near future. Thus, such seemingly indivisible phenomena as psychical processes (e.g., thought processes) can be successfully expressed in the neurophysiological terms of biological functioning of the brain. The next and final step would consist in explaining, in neurophysiological and cyber-netic terms, the very "core" of the psyche, i.e., the sense of one's own consciousness, of one's "ego" (in this book, I presented my own conception of the essence and the neurophysiological correlate of self-consciousness). Not much room has been left for the spirit in its immanence, as an autonomous being of equal standing with matter. The concept of the spirit simply turns out to be a void (a hypostasis) that does not find any correlate in the world. This is equivalent to a "semantic disintegration" of the concept, its total dissolution in a semantic void, as it corresponds neither to anything "real" nor even to anything sensible. Does it mean that materialism has finally been triumphant?

No, because there is also the other side of the coin. For, as biology (neurophysiology) deprived the spirit of its absolute attributes, so did physics with its eternal antagonist matter. Science instigated considerable changes in the meaning of the concept of "matter." More, the concept became "un-defined," decomposed into other concepts and thus practically devoid of its meaning that was "ceded" to the surrounding concepts. In commonsensical understanding (that is the starting point for philosophy), matter is something tangible, located and extending in space, endowed with weight, consistency, and color, something that can be transformed in various ways, but never destroyed. All these attributes of matter disappear in theoretical physics. The essence of the concept, so well known and clear, becomes very mysterious and exotic. Matter may disappear by transforming itself into energy, according to the famous Einstein's formula $E = mc^2$. In quantum mechanics, the position of an elementary particle, e.g., an electron, in space can be determined only approximately, according to Heisenberg's indeterminacy principle. Neither does it make sense to talk about the color of an atom or electron, when it does not emit the "color carrier"—an electromagnetic wave. The matter particles occasionally behave not like particles, but like waves: an electron can, for instance, pass through two slits in a diffraction grating (so, it can be located at two different places at one time!). Generally speaking, an electron is not located (before the act of measuring its position that is essentially a mental phenomenon, relativized with respect to the observer) in any particular place, but it is rather "distributed" in space. The spatial distribution of its position, corresponding to the probability of finding the electron in a given place in space, is described by the so-called wave function. We are unable, however, to univocally decide, whether a wave corresponding to this function is a material phenomenon, or rather a construct generated by our mind, and thus, we are unable to indicate a clear-cut border between reality and the content of the human consciousness! According to the general theory of relativity, matter is not only a material filling in flat space, but it shapes the space as well, by curving it proportionately to its mass. Recently, it has been more and more often claimed that the elements of the "objective" reality (e.g., matter and space, elementary particles and interactions) do not exist in the absolute sense, independently of each other, but rather their existence is mutually co-conditioned (and thus, they "mean"— in a sense—by connotation, just like concepts!). Quantum mechanics, in turn, suggests (in its modern interpretations) that all elementary particles that have ever been in any contact with each other, remain "correlated" with each other until now (the properties of some particles, e.g., spins, are not independent of the properties of other particles) and they constitute, in principle, one huge and indivisible network, and therefore, they must not be considered separately (that is the foundation of the methodology of modern physics, expressed in the following formula: first, examine system elements in isolation, only then, put them together!). In this muddle of absurdity and exotics, we are unable to recognize our old familiar matter that we learned about on the basis of a stone kept in the hand, and that was the subject of philosophy for over two thousand years. This seemingly clear and obvious concept passed through our fingers when we took a closer look at it.

One faces similar problems in the case of other elements of physical reality, such as space and time. According to the special theory of relativity, the rate of time flow and the length of a distance in space depend on the relative speed of the observer. The unidirectional time arrow and the correlated tendency of entropy to grow are to a large extent subjective (they are absent, for instance, from the Newtonian mechanics, the theory of relativity or quantum mechanics!) as distinguishing (more or less "complex") macroscopic states (carried out within thermodynamics) seems to be in principle a procedure relative with respect to our mind. A good example is provided by living organisms (thermodynamic phenomena by their nature) that are not "distinguished" in any special manner from the point of view of single atoms. Thus, it is impossible in physics to determine a border between a "real" element of our world and a purely mathematical model or a conceptual figment of our mind. The drive toward objectivization of the (view of the) world, so characteristic for natural sciences, paradoxically resulted in its progressive subjectivization.

When analyzing the impact of natural sciences on the problem of the relation between spirit and matter, we are surprised to face a kind of vicious circle. In this circle, spirit and matter co-condition each other, like two palms that draw each other on the famous Escher's image (or like two persons who exist only in each other's dream): matter generates consciousness, while consciousness determines the essence of what we call matter. In neurophysiology, self-consciousness turns out to be a product of matter (or a neural network in the brain), while in physics, matter seems to be more and more shaped by mathematical and conceptual constructs of our mind. The categories describing the so-called material world seem to be derived so much from the world itself, as from the human mental sphere, from the manner of integrating and processing sensory stimuli by our brain, conditioned by biological evolution. It may turn out that it does not make sense to choose between materialism and idealism, because we are unable to show a substantial or essential difference between matter and spirit (consciousness), and all the more so, since both concepts seem more and more clearly to be just void names that do not find any univocal and absolute correlates. Finally, we will probably conclude that none of the beings can be primary, as in one being indispensable for the other to come into existence (or at least to exist in a form that is known and accessible to us). If we come to such a conclusion, then probably the biggest problem of the "queen of sciences" will be solved (or rather considered to be absurd), not as a result of two thousand years of the development of philosophy, but thanks to the development of physics and biology in the twentieth century.

The relativity and immanent uncertainty of our knowledge—presented above by the example of spirit and matter—is a fundamental and universal feature that turns any dreams of an absolute truth into a utopia. One can utter only truths that are relativized to their semantic context, with reservations indicating (arbitrary) assumptions that have to be made for the truths to be considered as relatively valid. However, everyone who thinks about the essence of the world feels at one time tempted to try to formulate explicitly what we "really" know about the totality of phenomena available to us. I did yield to such temptation once, and here are the conclusions I arrived at when trying to determine "What do we know?":

1

Something undoubtedly exists in some way. There exists the so-called external world and the so-called content of our consciousness, irrespective of their nature, their mutual relations, and of the fact which is primary and which is secondary with respect to the other. Let us assume that this is our definition of existence.

2

We know not whether the existing something must exist or just may do so. We know not why it exists in this particular manner. We know not whether the concept of nonexistence (nonentity) makes sense at all.

3

We may analyze the entire available content of what exists and attempt to get closer to understanding the nature of existence in this manner. We may correlate various existing objects with each other and define the relative significance (essence, sense) of one object with respect to other objects. We may define one object by means of other objects. For instance, we may examine mutual interrelations between what we preliminarily know as matter (the physical world, the external reality), spirit (psyche, consciousness), living creatures, general beings, mathematical ideas, etc.

4

We cannot determine any "absolute" status of that which exists. As we analyze existence from inside, we cannot by principle investigate the ulti-

mate essence of the universe, for we lack an "external gauge" to use for this purpose. Therefore, we are doomed to have relative knowledge, as we by principle lack concepts to work out anything better.

5

Thus, we are only attempting to find, by means of the available conceptual apparatus, a possibly simple and universal description that would order in a possibly effective manner the whole of existence accessible to us. The criterion of description adequacy should be constituted by its usefulness, coherence, and heuristic fertility.

6

For the above reasons, it is convenient to propose a hypothesis (impossible to be demonstrated in any absolute manner) about the existence of the external material world. This concept allows us to segregate efficiently (and so, to order) the set of phenomena accessible to us (the totality of existence).

7

The external world is "external" with respect to something that is given to us directly, in a sense, and which can conveniently be termed as consciousness, without need to immediately determine its status.

8

Then, we may examine what relations obtain between the external world (matter) so understood and consciousness (mind). For this purpose, we take advantage of the "structure" of the world and mind, known to us in particular through the natural sciences.

9

Having carried out an adequate analysis, we reach—in the last instance—the above-discussed vicious circle, where spirit and matter are co-conditioned by each other.

10

What we can do is to treat spirit, matter, time, space, the sense of our own "ego" and all other phenomena as equally important concepts that signify by connotation and belong to a conceptual network that constitutes every-thing we can have access to. This is equivalent to admitting that our knowledge is only relative, and to resigning from any "absolute truth."

For an epistemological optimist desiring the absolute (in the form of a perfect and "final" mathematics, physics, philosophy, etc.), the above conclusions must seem to be very disappointing. What is the reason for our knowledge, including scientific knowledge, to be so uncertain and relative, as far as the most fundamental basis of its justification is concerned? As the reader must have already noticed, the uncertainty is caused by the gen-eral, connotative structure of the conceptual network that con-stitutes the substance of our consciousness and the cognitive apparatus and determines the nature, the justification, and the limitations of the worldview we have at our disposal. The claim that concepts (which is even more applicable to linguistic names ascribed to some of them) correspond to objects in (aspects of) the external world by denotation, i.e., by mutual one-to-one correspondence, is only a delusion of our mind (resulting from imperfections of self-cognition). To the contrary, a developing conceptual network "entwines"—to use the same analogy again—the objective "facts" of the world, just as a spider web may entwine the surface of a three-dimensional sculpture. When the mesh becomes finer, new objects/aspects of the world are fished out by the conceptual network or the sculpture surface is reflected in a more and more accurate manner by the spider web. The accuracy of the reflection, however, is never perfect. What's more, as the material of the spider web differs from the material of the stone sculpture, so is the substance of the conceptual network completely different from the substance of the world. The number of aspects of the external world reflected in its image shaped within a conceptual network, as well as the accuracy of their reflection depend on the degree (and direction) of the development of a given conceptual network. The external world appears in our mind as so vivid and rich that it may seem to us obvious that we perceive it as it actually is in its immanence. This is an utter delusion, however, since our mind "experiences" directly only stimulations of certain conceptual network fragments by stimuli (properly integrated and processed on their way) from sensory organs. Thus, there is no direct and absolute link between concepts and objects from/aspects of the world. A certain rough correspondence between the structure of the conceptual network and the structure of the world was established by trial and error, during biological evolution and during ontogenesis. As I have argued earlier, the representation of "the same" object in an external world, e.g., a roe deer, has a totally different form in the conceptual network of a tick and in that of a man (as the concept of a mother is different in an infant and in an adult). Chance plays a considerable role in the development and shaping of the final form of a conceptual network, both during biological evolution and during ontogenesis. If one considers the fact that the structure of a conceptual network is a derivative not only of the objective reality, but also of the structure of receptors and of the integration and association of signals flowing through them into the brain, one will understand that the external world is given to us only in a highly mediated (and distorted) form.

I believe that it resulted—at the moment, when man turned to philosophy (craving by its nature to reach the absolute truth)—in the appearance of an outrageous (from the human point of view) paradox that I call the evolutionary mind trap. Our mind (the cognitive apparatus) was shaped during the process of biological evolution within the objectively existing (whatever it could mean) world in such a manner (it could not have happened in any other manner) that it is not capable of proving even the existence of this world in an absolute manner. The only thing given to us as the so-called external world is a certain area of the conceptual network corresponding to it (at the neurophysiological level that belongs to the external world the view of this world is a fragment of the neural network and its "blind" ends—the neural connections to the receptors and effectors). This generated the temptation to consider the world as a figment (or at least a secondary element) of consciousness. On the other hand, the dependence of the mental content on the state of the body (especially the brain), observed centuries ago, was just a step away from concluding that consciousness is a product of matter. This is the origin of the above-discussed problem of the mutual relation between "spirit" and "matter." In its essence, the problem is insoluble in an absolute meaning, if it makes sense at all (which has been discussed above). Nevertheless, the "external world hypothesis" seems to be very useful in the operational sense, i.e., in the categorization and manipulation of the entire huge spectrum of phenomena accessible to us. And we can claim the objective world, as well as the subjective "ego," to exist only in this instrumental sense. In the last instance, both phenomena seem to be nothing more but differently shaped fragments of the conceptual network. The same applies to the question of existence or nonexistence of generalities (another great philosophical problem). The problem boils down to varying interpretations of the status of secondary concepts.[3] The undecidability or even nonsensicality of philosophical problems is a simple derivative of our evolutionarily shaped neurophysiology. The humanistic philosophy also failed to notice the fundamental fact that numerous immanently "human" properties of our psyche, our culture, and our view of the world were shaped by events, accidental in their nature, that have occurred during biological evolution.

If empirical data are not capable of providing us with absolutely certain knowledge, since our representation of the world created within the conceptual network contains as many "necessary" features of this world as it does contain accidental ailments of the network itself, perhaps the problem could be solved by the "pure reason" (the term introduced by Immanuel Kant) and its products, logic and mathematics, commonly considered to be universally true? Couldn't one feel the breath of the absolute at least in these domains? No. The problem is that the so-called pure reason—i.e., the seemingly most abstract reflection that is most removed from the world—finds its roots in the domain of the empirical. I am not going to insist on the literal *credo* of empiricism—*nihil in mente quod non prius in sensu*—as our mind is shaped not only by our senses, but also by the inborn integrative-associative mechanisms. These, in turn, are a derivative of the experience of our ancestors (again, from the empirical sphere!), but also of the brain material and of chance. Ultimately, our view of the external world has two components, the empirical one and the accidental one, yet the former allows the world to be mapped only approximately and imperfectly. The mind purified of the empirical component is left only with chance. Even such domains as logic and mathematics are empirical through and through, at least because they operate on objects and relations taken directly from the external world (or rather from sensory organs), once they have been abstracted and sublimated to purify them of certain properties. But they have no contact whatsoever with the ideal world of Platonic mathematical ideas (nothing like that exists, in my opinion).

Our logic and mathematics, i.e., the one we have access to (not to say: the one that has been created by our mind) is based exclusively on the conceptual network we have at our disposal that is derived from the real physical world (its reflection has been shaped within this reality during ontogenesis and biological evolution), but with a considerable admixture of accidentality and arbitrariness. In short, logic has no absolute relation to the real world, as it operates—by definition—on objects and categories (sets) that are isolated in an absolute manner, while there are simply no such things in this world, continuous by nature and resisting clear-cut categorizations. The same applies to arithmetic operating on integers: it has nothing to count in the real world, as there are no ideally isolated and defined phenomena in this world. The swarm of various types of geometries created by humans—some of which seem to describe the structure of the time-space of the universe—was corrupted at its very origin with a cardinal error. Mathematicians, believing in the absolute coherence and perfection of their discipline, have been attempting to disguise or de-translate it. I am referring here, simplifying the matter, to the trivial fact that the infinite number of points along a section of a straight line multiplied by the zero dimensions of a point yield the length of 3 (in some units, e.g., centimeters) in one case, while it yields 5 (in the same units) in another case (the same contradiction was formulated in a slightly different manner by Zeno of Elea in his famous paradox of Achilles and the tortoise). In more general terms, this is an expression of the problem of infinity (in this case, of the infinite power of the *continuum*), concerning both infinitely large quantities and infinitely small (zero) quantities. The above concepts used by mathematicians so freely and light-heartedly do not have, in my opinion, anything to do with the material reality. More, they are self-contradictory (such attempts to "de-translate" this contradiction, as the so-called nonstandard analysis, leads only—in my opinion—to shoving it into a slightly different conceptual context, where it is harder to track down the identified contradiction).

I believe that there exists one noncontradictory and internally consistent mathematics, equivalent by necessity with the structure of the real world. I believe that this mathematics combines the discrete character of arithmetic and the continuity of geometry, reconciling in some way these seemingly contradictory properties. "Our" arithmetic and geometry are only imperfect reflections of two different aspects of the only possible mathematics in our conceptual network[4] (it is a different question, whether the most basic constitution of our mind will ever allow us to construct such mathematics within our conceptual network). I believe that the multiplicity of the existing mathematical (and logical) systems and the freedom to create such systems, based on various combinations of axioms, is a clear indication that we have not reached the true foundations of mathematics yet, and the multiplicity is a simple derivative of contradictions and incoherences that are hidden from us or that we are only partly aware of.

The self-contradiction of logic (created within our conceptual network) can be demonstrated by the following example. It is known that a set is one of the basic concepts of logic. A set is a certain category that contains a number of elements (the number may be equal to zero, in the case of an empty set, or it may be finite or infinite). A set can contain concrete objects (a set of red cars), abstract objects (a set of Joseph K's thoughts related to the court) or other sets (e.g., a set of sets that are not their own elements, as in the famous Russell's antinomy—see below). Let us consider the set of all sets. Its meaning (i.e., intension) seems to be absolutely clear and easy to think, just like a set of all persons present in a room. But, here we face a surprise. It turns out that the "set of all sets" is a self-contradictory, if not nonsensical, cluster of words (it does not designate any object that would make sense, and therefore its extension does not exist in an absolute manner). For, the theorem that a set of all sub-sets of each set must be larger than the set, has been proved beyond any doubt. If the set of all subsets of the set of all sets is larger than the set of all sets, then the latter cannot be a set of all sets, i.e., it cannot be itself. A similar paradox is contained in the commonly known Russell's antinomy that consists in the fact that contradiction is reached by following both the positive and negative answer to the question, whether the set of all sets not being their own members is its own member. In the famous proof, Gödel constructed a correlate of the sentence "it is hereby proved that this entire sentence cannot be proven" in the system of mathematical theorems that has led to the debacle of the myth of com-

pleteness and coherence of mathematics. For me, what constitutes a catastrophe for logicians, destroying the supposed perfection of their domain, is just an expression of the connotative and
continuous ("analogue") structure of the conceptual network.

Generally speaking, logic and mathematics cannot pretend
to be absolutely coherent and "true" since both are forms of language (the most precise and formalized ones, to be sure). Language names do not refer directly (contrary to what one would
think at first) to objects of the real world or of the world of Platonic ideas. They simply denote, as I have explained above, various concepts in a conceptual network, usually those most
clearly isolated and defined. Thus, concepts constitute the only
way, the only intermediate link that allows one to establish the
correspondence between the facts of the (real or Platonic) world
and language names. As both the correspondence between the
facts of the world and the concepts in the conceptual network,
and the correspondence between concepts and language names
are not perfect, the discrepancy between the external (or Platonic) world and its linguistic description must be even greater.
This applies both to science, to the sphere of common sense,
and, in particular, to philosophy. The human mind, of course,
shows a special predilection to use language. The predilection is
so strong that the mind is ready to absolutize language names
and their meanings. This fact finds its clear expression, for
instance, in the notorious production of empty names
(hypostases) within philosophy, or in the ascription of causative
power to words in magic. The early Wittgenstein (*Tractatus
Logico-Philosophicus*) went as far as to identify the structure of the
language with the structure of cognition, and even with the
structure of the world. However, this is nothing but a delusion,
caused by the inborn proclivity of the human brain to use language, while the proclivity results from the fact that language

was evolutionarily useful in developing primeval societies. Language names—by their quantized and discrete nature—often isolate and determine aspects of reality in an absolutely clear-cut manner, while these aspects are not so well defined due to the structure of the world that is continuous in its essence.

This applies first of all to general terms, but also to the names of individual entities. For instance, it is a totally arbitrary and subjective operation to group atoms constituting a given person (let us call him John) into a single object, clearly isolated from its environment. John is constantly exchanging matter with the environment—e.g., he is breathing, eating food, and sweating (not to mention other physiological functions)—and therefore the set of atoms constituting him is being constantly renewed. If John were to be dismantled gradually into atoms, by subtracting atom by atom, nobody would be able to determine the boundary between John and non-John. This applies also to John's consciousness that is a derivative of the functioning of his brain, and in the last instance of a certain complex movement of atoms. It is commonly assumed (due to the tradition of thought) that the facts of the world exist independently and autonomously of language names. I claim, however, that the so-called facts are constituted by our language (and the conceptual cognitive structures of our brain at their base) from the "chaos," or rather "non-sense," of sensory stimuli storming the brain. Whether we consider John to be a fact, or not, depends on the manner of integrating sensations received from receptors. Facts are categories of our mind, and not objectively existing categories of the world. Therefore, language uses names that do not have well-defined designations, be it in the external world, or in the world of Platonic mathematical ideas. I discussed this problem in greater detail in my previous books (published in Polish).

The fact-creating nature of language has an impact on our

culture, science, religion, and common sense, and it shapes the form and content of human thinking. The methodology that ensures a (relatively good) correspondence between linguistic (and conceptual) structures and the structures of the world is what differentiates exact and natural sciences, for instance, from philosophy and religion. The correspondence, however, is far from perfect, and this is precisely what differentiates the real science from frequently idealized ideas about science. The error of hypostasis consists here in a naive belief in an absolutely real and "clear-cut" existence of such objects described by science, as the orbit of an electron, the force of gravity or the biological species that are abstract and/or general concepts, as well as such objects as the planet Earth or a particular individual of a given biological species. The term "real" does not determine in this case the existence outside our psyche in general—which I am not going to negate—but only the existence in forms, categories, or even values that are produced by our consciousness. I do not share the extreme view of Kant that "things in themselves" are totally unknowable. If our conceptual categories had nothing to do with the real structures of the world, we would not be able survive as a biological species, and much less to achieve such spectacular success in the development of science and technology that we did achieve, after all. Yet, even a very high correspondence of the conceptual network structure to the external world structure is by no means equivalent to perfect correspondence (besides, the degrees of correspondence may differ considerably in the case of different aspects of the world).

The relative and nonobvious character of the most fundamental properties of human insight into the world can easily be demonstrated, for instance, by comparing the sense of vision in humans and in other animals inhabiting the Earth. Image perception is the basic feature of our vision. If we look at something, e.g., on the picture entitled *Mona Lisa*, we see at least a certain spatial distribution of colorful spots (that are integrated by our brain, on the basis of the inculcated cultural patterns, into a portrait of a mysteriously smiling woman in mourning). However, if we looked at the same picture through the eyes (and especially through the brain!) of a frog, it would appear as something totally different. A frog perceives primarily movements. Therefore, a frog would see the picture hanging on the wall as a uniform background, devoid of colors or even brightness (thus, the very statement that a frog sees a monotonous background is already an abuse). On the other hand, a fly flying against the background of the picture, hardly be noticed by us, would elicit a violent reaction in the brain of the frog, since it would be the only object perceived by the animal! The eye of the amphibian in question is similar to the human eye: it is sensitive to the same range of electromagnetic radiation and it focuses the image on the retina by means of the lens. Therefore, the same spatial distribution of light signals results in both cases in a similar spatial distribution of stimulations of light-sensitive cell at the bottom of the eyeball. The reason why the frog perceives something else than a human consists in a totally different manner of integration of signals coming from the retinas to their brains. To simplify the issue, the human brain has a stronger tendency to create images on the basis of *spatially* differentiated stimulation of light-sensitive cells in different areas of the retina, while the brain of a frog prefers to register *temporal* changes in the stimulation of particular light-sensitive cells. In other words, the human brain receives with high intensity the information like "this light-sensitive cell is stimulated, but its neighbor is not" (or the other way around), while the brain of a frog most willingly reacts to the message like "this light-sensitive cell was not stimulated a while ago, but it is now" (or the other way around).[5] The mechanisms of sensory data processing are so deeply embedded in the most fundamental categories of our thinking and world-viewing that we are inclined to take them for the attributes of the world itself.

With the frog, we share at least a very similar construction of the visual receptor—the eye. However, when humans are compared to insects, even this similarity disappears. Insects receive the range of colors between ultraviolet and orange, while the human color reception range is between violet and red. Moreover, their eyes are constructed in a totally different manner. Insects have compound eyes, composed of simple eyes, the so-called ommatidia. Finally, insects perceive light polarization that is imperceptible for us. Therefore, we can suspect that their view of the world is totally incomparable to ours (irrespective of the huge difference in the size of the brain and the degree of its complexity).

Even in the case of "normal" (i.e., human) vision, dominated by the perception of images, the process of grouping certain sets of stimuli into "facts" or "cognitive structures" is a derivative of the integration of these stimuli by our brain. We can imagine beings, whose description of their world lacks categories corresponding well to our concepts of space, time, or causality (but equipped with some other configurations of concepts that are totally inaccessible to us). Most probably, they would be less adapted to live in the real world and to cognize it. It cannot be ruled out, however, that they would be able to do so just as well as we do. Their logic would differ from ours. Even "reality" would mean something else for them. We are unable to find the most fundamental element of our mind that we would be allowed to indicate as an objective fact or aspect of the world.

Another example of the seeming obviousness of our forms of world perception can be drawn from humanistically oriented philosophy (that I generally dislike). As I have already mentioned above, the division of the totality of accessible phenomena into a conscious "ego" and the external world surrounding it is probably the most deeply rooted categorization in the human mind. Martin Heidegger proposed to assume the relation of "being in the world" as the most primary phenomenon, and to consider its components—i.e., both *ego* and the world—as secondary with respect to this relation. He assumed "being in time" as an equally primary relation. I consider propositions of this type to be conceptual games of insignificant heuristic value, yet my cognitive honesty forces me to admit that I would not be able to indicate any essential advantage of the traditional view over Heidegger's proposition, apart from utility, perhaps.

Hitherto, we have been talking about differences between various organisms in the *purposeful* shaping (during biological evolution) of the structure of receptors and integrative mechanisms in the brain. However, totally *accidental* factors may have fundamental impact on the properties of the human brain. I proposed earlier the thesis that so sharp a differentiation between spirit (ego, self-consciousness) and matter (the physical world, the external reality) in our psyche may be caused by a considerable spatial distance between the sensory cortex (parietal and temporal), where the view of the world is located, and the "decision-making" cortex (prefrontal), where the operating memory, the "free will" and the sense of one's own "ego" is located. As we remember, these areas of the brain cortex are connected through the "central brain tract" that cannot be fully functional due to the length of the connections (over 10 cm for each axon). The tract could be absent from the brain, if the embryogenesis of the brain took a different form in our ancestors (compare figure 3 on page 35 and the related considerations). In such a case, the communication between the view of the world and the *ego* would be much more efficient, and both centers would be much better integrated with each other, and thus our psyche would never develop the differentiation into the content of consciousness and the external world (or the differentiation would have a different, less "conflictogenic" form). If I am right, the entire philosophy should start with neurophysiology and it should end with neurophysiology.

Let us remember the already-discussed sensory nature (substance) of concepts (at least partly, as we should remember the associative component) and, in particular, of conceptual cognitive structures that give form to newly created concepts. For instance, the final form of visual sensations is such and no other in our subjective experience, because our brain treats signals received from the eyes as visual signals. If the auditory nerve leading to the ear were cut and connected to the visual nerve leading to the brain (the visual cortex), we would see a chaotically moving mosaic of spots (something like the white noise on the TV). An inverse operation (connecting the visual nerve leading from the eye to the auditory cortex) would lead to a subjective experience of the cacophony of sounds.

The mechanisms integrating receptor signals have an overwhelming (and still underestimated) impact on the essence, the possibilities, and the range of human thinking and cognition. The most fundamental forms and categories of our insight into the external world are derived directly from the manner of processing receptor signals by our brain. To visualize it well, let us consider once again, how would the "world" of insects (equipped with compound eyes) or bats (creating the view of the world primarily through echolocation) look like. Our manner of viewing the world is certainly not the only possible one. Yet, this is the foundation of the entire huge edifice of philosophy, with pretenses to the absolute, including its principal domains:

ontology and epistemology. As philosophy (or at least its major part) fails to recognize the fact that it is not talking about the absolute objective external world, but only about a subjective and relatively incidental "world" shaped in our heads, the philosophical reflection becomes essentially deprived of any sense. Construction of philosophical systems without knowledge of the basic principles of functioning of the human brain is like attempts at predicating about the nature of matter (four elements, etc.) before the establishment of modern physics. For the status at concepts, the structure of the conceptual network and the resultant nature of our worldview (determining our cognitive limitations) must constitute the unconditional point of departure for any reliable ontology and epistemology.

Let us use a simple example: the innate strong tendency to perceive straight lines may easily throw light upon our natural predilection for Euclidean geometry. The mentioned predilection of the human brain to isolate discrete objects (that are not so clearly isolated "in reality") from the more or less continuous image finds its reflection in the tendency to quantify the world, which, on the one hand, provides a foundation for the development of language (consisting, after all, of discrete names denoting objects), while, on the other hand, it creates the need to develop the concept of a number (counting objects), which is just a step away from arithmetic. The concept (category) of causality, or of temporal continuity in more general terms, is a result—at the neurophysiological level—of joining into one object of a spot moving across the retina in such a way that the spot occupies neighboring spatial positions in successive moments in time, instead of moving, for instance, in a chaotic manner across the retina, jumping from one area to another.

Therefore, it should be obvious that the differentiation of such cognitive forms and categories as time, space, the causal relation or a discrete object (and, in principle, of all other concepts and categories) belongs totally to the domain of the perception of stimuli by our brain. All the integrative and associative mechanisms were, of course, given to us by the biological evolution that ensured that our categories and concepts should not be totally abstracted from the external world, and that "something" in this world should correspond to them. However, we will never be sure what it is, and no philosophy is ever going to help us with this problem.

Science, thanks to its methodology, ensures a much better correspondence of conceptual maps of its particular domains to various aspects of the external world than in the case of philosophy. However, science will never allow us to reach the ultimate and absolute essence of the world either. The reason is simple and trivial, in the light of the above-presented discussion. Science is only a fragment of a "collective conceptual network," and therefore it cannot pretend to correspond perfectly to the world structures (the further away from our everyday experience the aspects of reality examined by science tend to be, the greater the problems we face with the coherence of various domains of the conceptual network and their correspondence to reality, which is illustrated by quantum mechanics or the general relativity theory). Concepts are much better defined in science than in philosophy or in the sphere of common sense, but the difference is merely quantitative, and not qualitative. Therefore, I do not believe in a great unification of the entire science. Our insight into the world will remain fragmentary forever.

To conclude: the fact that concepts signify by connotation in a conceptual network has a hard-to-underestimate impact on the understanding of human consciousness, the degree of justification of our cognition and the status of the worldview created through the cognition. The connotative character of the conceptual network follows ineluctably from the structure of its physical carrier—the neural network. Since our mind has direct access only to the conceptual network, we can predicate about the external world, our self-consciousness, and its content only through the prism of that network. In the connotative conceptual network all senses and meanings exist only and exclusively in relation to other senses and meanings. Our propositions may only have a relative status, dependent on the context of the surrounding senses.[6] Concepts do not designate directly any objects outside the conceptual network, and their meaning is realized only with respect to other concepts. This leads ineluctably to the total impossibility of predicting truths that would be absolute in any manner and valid irrespective of the context. This is the funeral of the traditionally minded philosophy.

10.

CONCLUSIONS

We are nearing the end of our route from a single neuron to human self-consciousness. Following the main line of thought of this book, we tried to trace how a SYSTEM emerges from the functioning of a set of appropriately organized elements. On the way, we saw how nerve cells join into neural networks capable of self-modifying their own structures (i.e., capable of learning), how signals from receptors are processed by integrative and associative structures, how decisions are taken, to be implemented by their transcription into detailed directives concerning activation of particular effectors. We saw how instincts and the reward-punishment system direct the evolution of the neural network during ontogenesis, where emotions come from and why conscious free will is impossible. We learned the overpowering impact that the connotative nature of the neural network has on the essence of mental objects, i.e., concepts, and their derivatives, i.e., sensations, thoughts, memories.... We have understood the impact of the human brain functioning on the shape, the cognitive status, and the limitations of our worldview, presented within philosophy, logic, mathematics, or science. Finally, we have been shown the mechanism that allows self-consciousness and the psychic level of reality (the sphere of subjective mental states conditioned by self-consciousness) to emerge from the functioning of the brain at the biological (neurophysiological) level. Thus, we faced an attempt at explaining the very essence of humanity. Have we managed to understand it?

Yes and no. Even if the presented conception convinced us at the rational level that it is true and/or heuristically fertile, we are certainly not completely satisfied. Instead of receiving a purely instrumental, operating prescription for producing self-consciousness, we would prefer to see "with our own eyes" how consciousness emerges as a result of a certain specific type of *complexity* that characterizes the set of processes that consciousness is based on. One of the main messages of this book, however, is the thesis that this wish will remain impossible to fulfill forever. Our evolution-shaped brain is not able to attribute the same objective status to complexity that we ascribe to space or matter. Seeing the essence of self-consciousness is similar to seeing a four-dimensional cube or curved space-time. Our mind has no direct access to such phenomena. We can, however, approximate the phenomena by producing a series of operations (mathematical, cybernetic, conceptual) that lead to the construction of such objects, as well as by tracing the implications of their nature (e.g., their relations with other phenomena). We are unable to do anything more.

The relation of self-application (self-focusing) was proposed as the property that is at the root of the essence of self-consciousness. It is worth noting that this relation is identical with the conceptual structure of the famous logical and mathematical paradoxes—the liar's paradox, Russell's antinomy of classes, and Gödel's proof—that undermined the myth of coherence and consistency of logic and mathematics. If we accept the above proposition at its face value, we must accept the fact that we are all, in a sense, children of a paradox.

NOTES

Chapter 2: The Functions of a Neuron

1. In this book I will use the terms "consciousness" and "self-consciousness" interchangeably, since I believe that it is not logically possible for consciousness to exist without self-consciousness, which will be discussed later on.

2. Something like that happens during an epileptic seizure.

Chapter 3: Brain Structure and Function

1. I decided to disregard various cells that have auxiliary functions, e.g., glia cells.

2. Yet, indeterminism is not equivalent to free will—see further discussion.

3. There are three types of cones, characterized by the maximum response to electromagnetic radiation with the wavelength corresponding, respectively, to red, green, and blue color. However, a "red"

quantum of light may stimulate a "green" cone, although the probability of such stimulation is much lower than the excitation generated by a "green" quantum of radiation. Therefore, it is necessary to compare signals from several (tens) of cones of different "color" and deduct the "resultant" color, in order to univocally determine the color of the electromagnetic radiation that reaches a small area of the retina.

4. Descartes, for instance, believed that the soul inhabits the pineal gland.

Chapter 4: The General Structure of the Neural Network

1. The reverberating circuits cover the neural connections between the thalamus and the brain cortex, as well as the closed circuit of connections between different layers of the brain cortex that covers in succession: layer 4 (that receives signals from other brain parts), layer 2+3 (sending signals to the neighboring cortex areas), layer 5 and layer 6 (that send signals to other parts of the brain), and again layer 4 that closes the cycle. I believe that such circuits may have much in common with the genesis of self-consciousness discussed in one of the following chapters (this is due to the self-referential relation, the relation of focusing on itself that constitutes, in my opinion, the basis of self-consciousness). It is interesting that the above-described circuit is interrupted during sleep—layers 5 and 6 do not work—when consciousness is switched off.

2. I believe that the "future" colors of red and blue have very similar predispositions at the moment of birth, and they are practically indistinguishable (both in the neurophysiological sense and in the mental sense). Their differentiation is due to the association of signals from "red" and "blue" receptors with different sets of objects, established within the neural network as experience is gathered during ontogenesis. In my opinion, if an infant were presented only with objects (including faces, the sky and anything else) with reversed blue

and red colors, he/she would perceive the colors in a reversed manner (if subjective sensations are at all comparable). Such reversal would not be possible in the case of images and sounds, as the manner of their integration—already within the receptors, and later, in the sensory brain cortex—is totally different.

Chapter 5: Instincts, Emotions, Free Will

1. I believe that it is not "bare" genes that are the subject of evolution—as Dawkins proposed—but self-replicating systems, consisting—in the simplest cases—of nucleic acids and proteins. However, I will use the commonly known metaphor of a "selfish gene," since biological evolution is not the main subject of this book.

2. As far as I know, the term "reward system" is commonly used, but I believe that the name "reward-punishment system" is more adequate, which I will discuss in a moment.

3. It is easy to falsify, in my opinion.

4. I am referring here to a psychic correlate, i.e., the subjective "content" of emotions, and not to their generation that proceeds in the subcortical centers.

5. As a matter of fact, it has also been proposed that sensory stimuli only modulate and channel the spontaneous brain activity that is chaotic by nature.

Chapter 6: The Nature of Mental Objects

1. One can just as well use any other term instead of that of "concept": e.g., a "psychon" as a unit of the psyche, and talk about a "psychonal network." I do not like such neologisms. "Concept" is an accurate word because I focus especially on our cognitive abilities, i.e., on the type and degree of correspondence between the conceptual network and the external world.

2. Just as in the case of a good photograph, one can disregard the fact that it consists of emulsion grains and consider the spots of colors on the photo to be at least potentially continuous.

3. This is a partial sensation at the lowest level of integration that cannot exist autonomously, since a red spot must have some shape, an extend in space, etc. However, to understand and visualize the red color, one needs adequate fragments of the conceptual network, and in the final instance, the entire network is necessary.

4. Therefore, there is nothing like *qualia*, or beings imagined by philosophers and supposed to carry qualitatively different features of particular sensations.

5. Of course, I use the term "center" rather in the functional, and not the structural sense.

6. He did not have any contact with language until he was a teenager.

7. Not only single words, but also entire phrases, sentences, or even larger text segments may function as correlates of concepts.

Chapter 7: The Rise and Essence of Self-Consciousness

1. The liar's paradox consists in the fact that the enunciation "I am lying now" ("Hereby, I am lying") cannot be true, neither can it be false.

2. Although the structure of the thalamus (and the brain trunk) in the brain can play an important role in generating consciousness, which I have discussed above.

Chapter 8: Artificial Intelligence

1. In principle, although beyond the technological possibilities available currently or in the near future.

2. I do not believe it is. as it is possible to prepare a set of utterly

mechanical answers for each *finite* set of questions. This also applies to sets of *sequences* of interconnected questions and answers. Therefore, the Turing test would be conclusive only for an *infinite* set of questions, which deprives it of any practical sense. Moreover, the Turing test is based only on a linguistic expression (manifestation) of consciousness, while consciousness can express itself through facial expression, actions, the functioning in society, etc.

3. Thus, its intelligence would have to be *intentional*.

Chapter 9: The Cognitive Limitations of Humans

1. They can only be shown to lack aprioric justification.

2. Or two quarks: u and d that constitute protons and neutrons, and an electron.

3. I feel sad, when I think about all these philosophical treatises on the nature of generalities, while the entire problem disappears, when one realizes that they are nothing else but a psychic correlate of certain associative structures in our brain.

4. I will use a simple example of an imaginary solution, while making a qualification that I have no idea, whether it has any real sense. Let us imagine that the problem of infinite quantities can be solved by indicating that there exists the shortest possible distance in space (and time). Let us call it b. It does not mean that space is discontinuous and consists of discrete balls with diameter b (and time consists of moments that are b/c long, where c is the speed of light). When going down to smaller and smaller scales, we are simply not going to reach a distance smaller than b (as we are not going to exceed the speed of light, when increasing the speed gradually, according to the special theory of relativity). In the scales of the order of b, distances would not sum up additively. For instance, b + b would be smaller than 2b (just like great speeds do not sum up additively in the theory of relativity). Let us assume that the classical ("apparent") distance is l, while the hypothetical "true" distance is d, and let us assume that, for

instance, $d = 1 + b^2(b + 1)$. Then, for the classical distance (1) equal to 0, the true distance d would be equal to the smallest possible distance b. On the other hand, in the scales much larger than b, the apparent distance would be almost ideally equal to the true distance. This would explain why both distances seem identical in our macroscopic reality. The above solution is, in a sense, equivalent to the interdependence between the degree of space curvature and the scale of magnitude. Space is (approximately, due to the general relativity theory) flat in macroscale, while it is infinitely curved (zero radius of curvature) in the scale of the b order. The value b could correspond to the so-called Planck length, equal approximately to 10^{-35} m.

Such solution would have unprecedented consequences for physics. For instance, the length of the light wave (λ) could not be smaller than b, and therefore, there would exist a maximum possible photon energy ($E = hc/\lambda$, where h is the Planck constant). Potential wells would not approach infinite values near their centers (the value of a potential, e.g., gravitational or electric, is inversely proportionate to the square of the distance) and there would be no need to use the renormalization procedure to avoid the infinite mass of the electron, obtained within the QED. No singularities would appear within the general theory of relativity. I hope mathematicians would not bridle up in the face of my "mathematical fiction." I do not mean to propose a concrete solution, but only to demonstrate—in an imaginary example—how deep changes in the most fundamental structure of the currently known mathematics and physics would be necessary for a really radical revolution (if it is possible at all) in our understanding of the real world to occur.

5. The difference between the frog and man is not so dramatic, as it could be suggested by this simplified description. Humans have also a certain predilection to perceive movement, especially in the dark and at the edges of the visual field.

6. One may wonder at the fact that relativity and co-conditioning appear more and more often in such a traditionally absolutistic and "objective" science like physics.

SUGGESTED FURTHER READING

Abu-Mostafa, Yaser S. "Machines That Learn from Hints." *Scientific American* (April 1995): 64–69.

Barrow, John D. *Pi in the Sky: Counting, Thinking, and Being.* Back Bay Books, 1993.

Beardsley, Tim. "The Machinery of Thought." *Scientific American* (August 1997): 78–83.

Bower, James M. and Lawrence M. Parsons. "Rethinking the 'Lesser Brain.'" *Scientific American* (August 2003): 50–57.

Cairns-Smith, A. Graham. *Evolving the Mind: On the Nature of Matter and the Origin of Consciousness.* Cambridge University Press, 1998.

Calvin, William H. *How Brains Think: Evolving Intelligence, Then and Now.* Basic Books, 1997.

———. "The Emergence of Intelligence." *Scientific American.* (October 1994): 100–107.

Chalmers, David J. "The Puzzle of Conscious Experience." *Scientific American.* (December 1995): 62–68.

Crick, Francis. *The Astonishing Hypothesis: The Scientific Search for the Soul.* New York, Scribner, 1995.

Crick, Francis and Christof Koch. "The Problem of Consciousness."

Scientific American. (September 1992): 152–59.

Damasio, Antonio R. *The Feeling of What Happens: Body and Emotion in the Making of Consciousness.* Harcourt, 1999.

———. "How the Brain Creates the Mind." *Scientific American.* (December 1999): 112–17.

———. "Remembering When." *Scientific American.* (September 2002): 66–73.

Dennett, Daniel C. *Kinds of Minds: Toward an Understanding of Consciousness.* Basic Books, 1997.

Gazzaniga, Michael S. "The Split Brain Revisited." *Scientific American.* (July 1998): 51–55.

Greenfield, Susan. *The Human Brain: A Guided Tour.* Basic Books, 1998.

Grillner, Sten. "Neural Networks for Vertebrate Locomotion." *Scientific American.* (January 1996): 64–69.

Ingram, Jay. *The Burning House: Unlocking the Mysteries of the Brain.* Penguin Books, 1996.

LeDoux, Joseph E. "Emotion, Memory and the Brain." *Scientific American.* (June 1994): 50–54.

Loftus, Elizabeth F. "Creating False Memories." *Scientific American.* (September 1997): 70–75.

Logothetis, Nikos K. "Vision: A Window into Consciousness." *Scientific American.* (Special Edition, November 2006): 4–11.

Nestler, Eric J. and Robert C. Malenka. "The Addicted Brain." *Scientific American.* (March 2004): 78–85.

Penrose, Roger. *The Emperor's New Mind: Concerning Computers, Minds, and the Laws of Physics.* Oxford University Press, 2002.

Raichle, Marcus E. "Visualizing the Mind." *Scientific American.* (April 1994): 58–64.

Siegel, Jerome M. "Why We Sleep." *Scientific American.* (November 2003): 91–97.

Smith, David V. and Robert F. Margolskee. "Making Sense of Taste." *Scientific American.* (March 2001): 32–39.

Ross, Philip. "Mind Readers." *Scientific American.* (September 2003): 74–77.

Weinberger, Norman M. "Music and the Brain." *Scientific American.* (October 2004): 89–95.

Winson, Jonathan. "The Meaning of Dreams." *Scientific American.* (Special Issue, 2002): 54–61.